Studies in Systems, Decision and Control

Volume 88

Series editor

Janusz Kacprzyk, Polish Academy of Sciences, Warsaw, Poland
e-mail: kacprzyk@ibspan.waw.pl

About this Series

The series "Studies in Systems, Decision and Control" (SSDC) covers both new developments and advances, as well as the state of the art, in the various areas of broadly perceived systems, decision making and control- quickly, up to date and with a high quality. The intent is to cover the theory, applications, and perspectives on the state of the art and future developments relevant to systems, decision making, control, complex processes and related areas, as embedded in the fields of engineering, computer science, physics, economics, social and life sciences, as well as the paradigms and methodologies behind them. The series contains monographs, textbooks, lecture notes and edited volumes in systems, decision making and control spanning the areas of Cyber-Physical Systems, Autonomous Systems, Sensor Networks, Control Systems, Energy Systems, Automotive Systems, Biological Systems, Vehicular Networking and Connected Vehicles, Aerospace Systems, Automation, Manufacturing, Smart Grids, Nonlinear Systems, Power Systems, Robotics, Social Systems, Economic Systems and other. Of particular value to both the contributors and the readership are the short publication timeframe and the world-wide distribution and exposure which enable both a wide and rapid dissemination of research output.

More information about this series at http://www.springer.com/series/13304

Krishnendu Mukherjee

Supplier Selection

An MCDA-Based Approach

Springer

Krishnendu Mukherjee
Mechanical Engineering
University of Engineering and Management
Jaipur, Rajasthan
India

ISSN 2198-4182 ISSN 2198-4190 (electronic)
Studies in Systems, Decision and Control
ISBN 978-81-322-3698-6 ISBN 978-81-322-3700-6 (eBook)
DOI 10.1007/978-81-322-3700-6

Library of Congress Control Number: 2016963662

Printed on acid-free paper

This Springer imprint is published by Springer Nature
The registered company is Springer (India) Pvt. Ltd.
The registered company address is: 7th Floor, Vijaya Building, 17 Barakhamba Road, New Delhi 110 001, India

I would like to thank my mentors—Prabir Kumar Bose, Ardhendu Bhattacharayya, and Bijon Sarkar—who have taught me the value of decision making and its effect on human life. I am also grateful to my parents for their love and encouragement. Specifically, I am indebted to my mother for her continuous support to complete this work.

Finally, this entire work is dedicated to Lord Krishna and to my mother. Without their blessing it would be difficult for me to give my best during my odd days.

Preface

Supplier selection is not a mere clerical process. It is a product-specific multi-criteria-based optimization problem. Judicious selection of suppliers could minimize procurement cost, enhance quality and reliability of supply, and thereby enhance profit margin of the company by minimizing upstream supply chain risk. Today, supply chain surplus of a company depends considerably on proper selection of its suppliers. Success of Apple Industry is a good indication of that. A decision maker has to trade off tangible and intangible criteria to select the best supplier or suppliers for the focal company. Several methods are available for selecting the best ones. Among these methods, the analytic hierarchy process (AHP) and the analytic network process (ANP) are the most used for their computational simplicity and consistency. This book encompasses several criteria and methods for supplier selection in a systematic way based on extensive literature review from 1998 to 2012 and includes illustrative case studies and examples for interested researchers.

The purpose of this book is to present a comprehensive review of the latest research and development trends at the international level for modeling and optimization of the supplier selection process for different industrial sectors. It is targeted to serve two audiences: the MBA and Ph.D. student interested in procurement and the practitioner who wishes to gain a deeper understanding of procurement analysis with multi-criteria-based decision tools to avoid upstream risks to get better supply chain visibility. This book is expected to serve as a ready reference for supplier selection criteria and various supplier evaluation methods for forward, reverse, and mass-customized supply chain. This book also encompasses strategic sourcing in detail for forward, reverse, and sustainable mass-customized supply chain.

Chapter 1 introduces the basic concept of multi-criteria decision-making process. It also introduces fundamental idea and steps of decision-making process. In brief, Chap. 1 gives fundamental idea of AHP, type of scale, prioritization methods, problem of rank reversal, validation of AHP, fuzzy hierarchical TOPSIS, VIKOR, uncertainty analysis with multi-criteria decision analysis tools, cascaded fuzzy inference system, and intuitionistic fuzzy set. Chapter 2 reviews supplier selection methods for traditional, reverse, and mass-customized supply chain. Chapter 3 mentions steps to develop mathematical model for supplier selection, evaluation, and order allocation problem. Chapter 4 discusses in detail about strategic sourcing. It includes several mathematical models with illustrative case studies. Chapter 5 discusses mainly on the limitations of fuzzy analytic hierarchy process and consistency of decision to aware researchers about certain limitations of fuzzy analytic hierarchy process. All mathematical models are solved with either MATLAB or R or LINGO. In the appendix, some computer code written in MATLAB, R, and VB.NET is also included for the interested reader.

I would like to give sincere thanks to Prof. T.L. Saaty, the originator of AHP and ANP method, and distinguished University Professor of Business Analytics and Operations Department of Joseph M. Katz Graduate School of Business, University of Pittsburgh, USA, for sharing his notes, published and unpublished research work on AHP and ANP. Without his continuous support and inspiration, it was difficult for me to know AHP and ANP in detail. My special thanks are due to Prof. Rozann Whitaker Saaty of the Creative Decisions Foundation of Pittsburgh, USA, for sharing her research work on AHP and ANP. I wish to thank Professor Timothy J. Ross of Civil Engineering Department of University of New Mexico; Professor Ahti A. Salo of Mathematics and Systems Analysis Department of Aalto University School of Science, Finland; and Dr. Matteo Brunelli, postdoctoral researcher of Systems Analysis Laboratory of Aalto University, Finland, for their wholehearted support and great encouragement in producing this book. I wish to thank various researchers and my students for their support to complete my work. I am also indebted to editorial team at Springer, for their support, assistance, and guidance on this book.

While every attempt has been made to ensure that no errors (printing or otherwise) may be present, the possibility of their existence is expected. As the saying goes, to err is human. I would be grateful to the readers if these errors are pointed out to me. Suggestions for further improvement of the book would be thankfully acknowledged. If any fact, data, figure, concept, etc, resemble with any published and/or unpublished work, then it is unintentional and I would remain grateful to readers if they kindly acknowledge it.

Jaipur Krishnendu Mukherjee

Contents

About the Author

Krishnendu Mukherjee is an operations research scientist with the Operations Research Machine Learning & Analytics Experts (ORMAE), where he has been dealing with several projects for multi-national companies in India and abroad. He did his Bachelor of Engineering in mechanical from Jadavpur University and Master of Engineering from BITS Pilani. He also worked as invited reviewer of IJPE, EJOR, JORS, IJAHP, etc. In 2004, he got opportunity to design and develop innovative carpet backing machine at IICT, Bhadohi, as one of the core team member. In 2014, he received US copyright for developing computer code to facilitate selection and evaluation of n-number of suppliers in fuzzy environment. In 2015, he introduced new concept of postponement strategy to prevent upstream supply chain risk. He has published 15 papers in international journals/conferences. His areas of research include multi-criteria decision analysis, fuzzy set theory, mixed-integer programming, linear programming, constrained programming, non-linear programming, optimization, supply chain management, sustainability, procurement analysis, vehicle routing, airlines scheduling, and tankers scheduling. He has also completed his Ph.D. work including the preparation of final draft of Ph.D. thesis at Jadavpur University. This book encompasses extended work of his Ph.D. thesis. He previously worked with National Institute of Technology Silchar (former REC); Ministry of Manpower, Oman; Indian Institute of Carpet Technology, Ministry of Textile, Govt. of India; Heritage Institute of Technology, Kolkata; MAHE, Dubai, etc.

Notations

AHP	Analytic hierarchy process
ANP	Analytic network process
ATO	Assembly to order
CA	Cluster analysis
CBR	Case-based reasoning
CI	Consistency index
CODP	Customer order decoupling point
DEA	Data envelopment analysis
EM	Eigenvalue method
EPAs	Environmental Protection Agencies
ETO	Engineer to order
FAHP	Fuzzy analytic hierarchy process
FANP	Fuzzy analytic network process
FNIS	Fuzzy negative ideal solution
FPIS	Fuzzy positive ideal solution
GA	Genetic algorithm
GDM	Group decision-making method
GHGs	Greenhouse gases
GP	Goal programming
GPD	Green product design
IFS	Intuitionistic fuzzy set
LLSM	Logarithmic least square method
MADM	Multi-attribute decision making
MAUT	Multi-attribute utility theory
MCDA	Multi-criteria decision analysis
MCDM	Multi-criteria decision making
MEFAHP	Modified extent fuzzy AHP
MEFAHP-GA	Modified extent fuzzy AHP and GA
MIP	Mixed-integer programming
MODM	Multi-objective decision making

MOGA	Multi-objective genetic algorithm
MTO	Make to order
MTS	Make to stock
NIS	Negative ideal solution
PFA	Product family architecture
PIS	Positive ideal solution
SCV	Supply chain visibility
TFN	Triangular fuzzy number
TOPSIS	Technique for order preference by similarity to ideal solution
TVP	Total value of purchase
TVRP	Total value of reliable purchase
VIKOR	VlseKriterijumska Optimizacija I Kompromisno Resenje in Serbian

List of Figures

List of Tables

Chapter 1
Overview

1.1 Introduction

"People make three general types of judgments to express importance, preference, or likelihood and use them to choose the best among alternatives in the presence of environmental, social, political, and other influences. They base these judgments on knowledge in memory or from analyzing benefits, costs, and risks. From past knowledge, we sometimes can develop standards of excellence and poorness and use them to rate the alternatives one at a time".

Thomas L Saaty
Distinguished Professor of Business Analytics and Operations,
Joseph M. Katz Graduate School of Business,
University of Pittsburgh, USA, (INTERFACES, 1994, Vol. 24, pp. 19–43, with kind permission from Springer Publishing Company)

The quote, above, addresses three types of decision in addressing large, complex, and most complicated decision-making process of human brain. Every decision-making process encompasses three key characteristics, namely complexity, credibility, and uncertainty. We do take decision even in complete uncertain situation to bring more credibility and thereby to reduce the complexity. Thus, judicious decision making is an art and the seed of successful life. The word 'decision' is originated from the latin word dēcīsiō, means cutting off. It means to select the best from available alternatives or converging to a conclusion. Every decision has the potential to change the world. Thus, every people must learn the art of decision making. Our success is the yield of our decision. Usually, every decision encompasses spiritual quotient (SQ), emotional quotient (EQ), and intelligent quotient (IQ). Person-to-person priority of each quotient varies. A spiritual person usually gives more priority to spiritual quotient, an emotional person usually gives more priority to emotional quotient, and an intelligent person usually gives more priority to intelligent quotient. It is expected to have different decisions of two

K. Mukherjee, *Supplier Selection*, Studies in Systems, Decision and Control 88,
DOI 10.1007/978-81-322-3700-6_1

persons in same environment because of infinite combinations of IQ, EQ, and SQ. Thus, every decision is the unique identity of each person and because of that we receive success or failure. Our decision is also influenced by political, social, and other influences. Sometimes one wrong decision triggers series of disasters. We learn from our past and use our past experience to take right decision. In Indian mythology, Lord Krishna asked Arjuna to kill all his enemies including his relatives and gurus those who opposed to establish dharma. Every word of Lord Krishna shows that the right decision at right place is always expected to avoid disaster like the gruesome war of Kurukshetra. The inherent characteristic of every human being is to move from less favorable state to more favorable state. This eternal journey demands right decision and consistent effort. Every good decision of human being depends on insight, evaluation of fact and external stimuli, imagination, ability to forecast future effect from present and past data, and language (Saaty and Shih 2009). Our every work demands good decision and priority. Thus, the art of decision making is truly justified to have better life. Usually, decision-making process consists of the following eight steps:

1. *Problem definition*: Define problem with root causes, constraints, or limitations.
2. *Need identification*: Elaborately explain the need to identify feasible solution space.
3. *Define goals*: What objective you are looking for to solve the problem?
4. *Select alternatives*: Alternatives are the paths to achieve predefined objective.
5. *Identify suitable criteria*: Which criteria are the best to achieve your goal or objective?
6. *Select appropriate decision-making tool*: A less complicated multi-criteria tool which gives error-free result to any problem is always preferred.
7. *Evaluate all alternatives against criteria*: Which alternatives are the best to achieve your goal? This assessment could be objective or subjective or combination of two to measure preference of selection of every alternative with respect to a criterion to achieve predefined objective.
8. *Validate result*: Always cross-check the result obtained from any decision-making process prior to its use.

The term 'problem' refers to the dissatisfaction or discontent or irritation perceived from some undesired result or effect of ongoing situation. Every human being and even insects move from less favorable state to more favorable state. Thus, making decision to survive in real life is a natural phenomenon. In 1980, a pragmatic method was proposed by **Saaty,** popularly known as analytic hierarchy process (AHP), to give hierarchical interpretation of problem with simple diagram. In 1974, **Altshuller** proposed **TRIZ (theory of solving inventive problems)** to solve problem in uncertain environment. Usually, our cognition power, pattern matching ability, associative memory, judicious judgment, ability to compare results, imagination, logical deduction, etc. help to solve complex problems of real life. Altshuller standardized the method of problem solving as follows:

1. Solve problem step-by-step in a systematical way.
2. Try to converge from broad solution space to achieve feasible optimal solution exact to or near to an ideal solution.
3. Try to produce reliable solution repeatedly.
4. Understand the body of inventive knowledge and use it judiciously.
5. Interact and discuss with inventors and explain the use of aforesaid method.

Saaty and Shih (2009) proposed the following steps to form hierarchy to solve problem:

1. Define the goal first.
2. Achieve objective of any problem in series. Highest level should be the goal of the problem.
3. Insert actors into a suitable level to achieve goal or objective.
4. All alternatives should be placed at bottom level.
5. Verify the hierarchic levels in forward and backward calculations.

Both Saaty and Altshuller proposed modular approach to solve complex problem even in uncertain environment in a systematic way. In decision making, structure is the first step to prepare the replica of the problem. It is the logical identification and grouping of similar things together to identify the flow of influence among them (Saaty and Shih 2009).

1.2 Characteristics and Classification of Criteria

Criteria are used to measure preference associated with each alternative with respect to goal. Criteria can also be defined w.r.t one or several preference scales related to the actual problem. The **scales** can be classified as continuous scale, discrete scale, ordinal scale, or cardinal scale (Jacquest Eghem Jr et al. 1989). Criteria should be non-redundant and relevant to the problem, and no. of criteria should be as small as possible (Baker et al. 2002). Selection of appropriate criteria is the cornerstone in decision-making process to select right alternative(s). Decision-maker's preference can be explained with the four binary relations introduced by (Jacquest Eghem Jr et al. 1989):

1. **Indifference situation**: It is denoted by a **I** b (i.e., a is indifferent to b). For example, two alternatives are equally important with respect to some criterion to a decision-maker.
2. **Preference situation**: It is denoted by a **P b** (i.e., a is strictly preferred to b). For example, one alternative is highly preferred with respect to one criterion to a decision-maker.
3. **Weak preference situation**: It is denoted by a **Q** b (i.e., degree of hesitation is associated with preference and indifference situation). For example, two

alternatives cannot be judged with respect to one criterion because of lack of certainty.

4. **Incomparability situation**: It is denoted by a **R b** (i.e., degree of hesitation is associated with a **P** b and b **P** a). For example, two alternatives cannot be compared with respect to one criterion because of ambiguity.

There may be other preference relations. Usually, MCDA methods are modeled with above four binary relations.

1.3 Classification of Decision Problem

According to Roy (1985), problem statements can be classified as follows:

1. ***Choice problem statement***: This type of problem assists the selection of one or several 'suitable action(s).'
2. ***Sorting problem statement***: This type of problem assists with group's various actions.
3. ***Rank problem statement***: This type of problem orders the actions w.r.t the predefined preference.

1.4 MCDA

Complex and conflicting real-life problems can be solved in simple but pragmatic way with multi-criteria decision analysis tools (MCDA). MCDA tools can be broadly categorized as **multi-attribute decision making (MADM)** and **multi-objective decision making (MODM)**. MADM is suitable for finite set of alternatives. MODM, on the other hand, is suitable for infinite number of alternatives. MCDA methodology can be considered as nonlinear recursive process which consists of four steps: (1) defining the problem, (2) identifying preference or criteria, (3) aggregating preferences, and (4) ordering alternatives (Guitouni and Martel 1998).

Opricovic and Tzeng (2004) proposed the following steps of MCDM:

1. Define suitable criteria to achieve goal(s).
2. Identify alternatives for achieving desired goal(s).
3. Evaluate each alternative w.r.t each criterion.
4. Use a suitable multi-criteria analysis tool or techniques.
5. Accept suitable alternative(s) to achieve goal(s).
6. If final solution is not feasible or not acceptable, then opt for next iteration until feasible solutions are not achieved.

Every human being does take decision rationally, irrationally, and non-rationally. In **rational decision**-making process, we select the best by evaluating all available alternatives. Often, we give priority based on our knowledge and past experiences. This approach is known as **non-rational decision**. We also express our personal like and dislike. This approach is known as **irrational decision** (Guitouni and Martel 1998).

Pairwise comparison is inherent characteristics of human being, and the majority of MCDA tools prefer to use pairwise comparison matrix to come to consensus through different methods of normalization and aggregation. MCDA is applicable to finite number of alternatives and it can be classified as follows:

1. The single-synthesizing criterion approach without incomparability (TOPSIS, MAVT, MAUT, AHP, EVAMIX, and UTA).
2. The outranking synthesizing approach.
3. The interactive local judgments with trial and error approach.

1.4.1 Analytic Hierarchy Process (AHP)

In AHP, palpable and non-palpable criteria are trade off to prepare a framework based on ratio scales from simple pairwise comparisons. (Schoner and Wedley 1989 as mentioned in Malcom Beynon 2002).

Analytic hierarchy process can easily be employed with the following steps:

Step 1: Define goal or objective.
Step 2: Define criteria and sub-criteria to accomplish goal.
Step 3: Use Saaty's nine-point preference scale to form pairwise comparison matrix. Let B is a $n \times n$ **pairwise comparison matrix**.

$$B = \begin{bmatrix} b_{11} & b_{21} \ldots & b_{1n} \\ b_{21} & b_{22} \ldots & b_{2n} \\ b_{n1} & b_{n2} \ldots & b_{nn} \end{bmatrix} \quad \text{where } b_{ij} = 1 \; \forall i = j$$

Step 4: Normalize the matrix with geometric mean as follows:

$$u_i = \frac{\left[\sum_{j=1}^{n} b_{ij}\right]^{1/n}}{\sum_{i=1}^{n}\left[\sum_{j=1}^{n} b_{ij}\right]^{1/n}} \quad \forall\, i,j = 1,2,3\ldots,n$$

Step 5: If R denotes n-dimensional column vector describing the sum of $R = [R_i]_{n\times 1} = Bu^T, = i = 1,2,\ldots n$

$$\text{where } Bu^T = \begin{bmatrix} 1 & b_{12}\ldots & b_{1n} \\ b_{21} & 1\ldots & b_{2n} \\ b_{n1} & b_{n2}\ldots & 1 \end{bmatrix} [u_1 u_2 \ldots u_n] = \begin{bmatrix} R_1 \\ R_2 \vdots \\ R_n \end{bmatrix}$$

Step 6: Saaty proposed the use of maximum **eigenvalue** $\lambda_{\max}$ to check consistency of judgment. The maximum eigenvalue $\lambda_{\max}$ can be determined as follows:

$$\lambda_{\max} = \frac{\sum_{i=1}^{n} Ru_i}{n} \; \forall i = 1, 2, 3 \ldots n$$

Step 7: Calculate **consistency index (CI)** as follows:

$$\mathrm{CI} = \frac{\lambda_{\max - n}}{n - 1}$$

Step 8: Calculate consistency ratio (CR) to check the validity of decision-making process.

$\mathrm{CR} = \frac{\mathrm{CI}}{\mathrm{RI}}$, where RI denotes average random index. Saaty proposed that CR value should be less than or equal to 0.1. Some researchers also mentioned that **random index (RI)** can be represented as follows: RI = $\{1.98(m - 2)/m\}$, where m is the size of the matrix.

1.4.2 *Types of Scale*

In multi-criteria decision analysis, every preference is associated with a suitable scale to evaluate or rank alternatives to achieve goal. For instance, A is two times better than B. Here, linguistic term 'two times better than' needs a scale to measure degree of preference. A scale is a triplet, encompasses a set of numbers, a set of objects, and mapping of objects to the number (Saaty 2004). There are different types of scale such as

1. **Nominal scale**: A number is assigned to each object. For example, queue in front of railway ticket reservation counter.
2. **Ordinal scale**: Numbers are assigned to each object to represent their order, increasing or decreasing.
3. **Interval scale**: For example, $Y = B - 30$, where Y is a dependent variable and B is an independent variable.
4. **Ratio scale**: For example, $K = BL$, where $B > 0$, and is a proportional constant.
5. **Absolute scale**: Number is used directly for pairwise comparison. It is commonly used in AHP.

Apart from above, there are eight different scales identified by Ishizaka and Labib (2011)—**linear scale** (Saaty 1977), **power scale** (Harker and Vargas 1987), **geometric scale** (Lootsma 1989), **logarithmic scale** (Ishizaka et al. 2010), **root square scale** (Harker and Vargas 1987), **asymptotical scale** (Dodd and Donegan 1995), **inverse linear scale** (Ma and Zheng 1991), and **balanced scale** (Salo and Hamalainen 1997). Usually, Saaty's scale and geometric scale are used. However, Saaty's scale is not like geometric scale which is considered as transitive scale (Dong et al. 2008). In 1994, Saaty mentioned categorically that the determination of parameter of geometric scale is difficult. Thus, Saaty's scale is must for AHP or ANP.

1.4.3 Prioritization Methods—EM or LLSM—Which One Is Better?

There are several prioritization methods available for multi-criteria decision analysis (Srdjevic 2005; Choo and Wedley 2004). Some of them are as follows:

1. Eigenvalue method (EM)
2. Logarithmic least square method (LLSM)

Selection of the best prioritization method is an open research issue irrespective of the most preferred eigenvalue method. Dong et al. (2008) proposed two algorithms to evaluate the performance of four scales and prioritization methods. Saaty (1990) proposed ten best reasons for using eigenvalue method and some of them are as follows:

(1) EM yields unique solution.
(2) Simplicity of EM.
(3) EM procedure is descriptive.
(4) EM does not consider statistical indices of bias.

1.4.4 Rank Reversal in AHP

In 1984, Belton and Gear identified rank reversal issues as the major concern for the application of AHP and further proposed to modify existing normalization method for classical AHP. Saaty and Vargas (1984) counter reply to the normalization method of Belton and Gear and declared rank reversal problem as natural phenomenon. Belton and Gear (1982) showed that the addition and deletion of alternative can change rank order if new alternative is neither strongly dominated by the least preferred alternative or strongly dominated by the most preferred alternative for every criterion. Wang and Elhag (2006) proposed that rank order of the alternatives cannot be changed if the priority of the new alternative did not influence the original local priority of every alternative under each criterion. They further proposed that if $B = (b_{ij})_{n\times n}$ is the pairwise comparison matrix, then after the addition of new alternative, comparison matrix becomes $B^1 = (b_{ij})_{(n+1)x(n+1)}$ and the priorities are represented by $W_B = [W_{1B}, W_{2B}, \ldots, W_{nB}]^T$ for the old pairwise comparison matrix and the new priorities are represented by $W_{B^1} = [W_{1B}, W_{2B}, \ldots W_{(n+1)B}]^T$. The necessary condition to preserve rank of alternatives after the addition of new alternative can be expressed as follows:

$$\sum_{i=1}^{n} W_{iB} = \sum_{i=1}^{n+1} W_{iB^1} = 1$$

Extensive literature review shows the following modified methods of AHP to tackle rank reversal problem:

1. **B–G modified AHP**, proposed by Belton and Gear (1982).
2. **Referenced AHP**, proposed by Schoner and Wedley (1989).
3. **Linking pin AHP**, proposed by Schoner et al. (1997).
4. **Multiplicative AHP**, proposed by Barzilai and Lootsma (1997).

Example The performance of two suppliers A and B is to be measured w.r.t two criteria price and quality. Equal priority is given to both criteria. Let us consider B is preferred to A. If another supplier C is introduced, whose performance is not better than A, then ranking will remain the same but their global or normalized priority value will change. If the performance of C is much better than B, then rank of B will change but rank of A remains the same.

A	*B*
0.46	0.54

$C \succ B \succ A$

A	*B*	*C*
0.37	0.47	0.17

$B \succ C \succ A$

A	*B*	*C*
0.13	0.36	0.5

1.4.5 Validation of AHP

In multi-criteria decision analysis, decision-makers prepare a mathematical model which is similar to given problem. Result obtained from the theoretical model does vary from the actual problem due to the assumptions considered to prepare mathematical models. The validity of mathematical model in decision-making process, thus, is highly required. Choice of numerical scale and method of prioritization highly influence the validity of decision-making process (Dong et al. 2008). Qureshi et al. (1999) proposed **VVS model** which consists of verification, validation, and sensitivity analysis to compare theoretical MCDA model with actual problem. They further proposed that to compare result, statistical inference approaches should be used. Interested reader can refer to work of Whitaker (2007) in this regard. Every real-life problem encompasses uncertainties. Usually, researchers prefer to use fuzzy AHP instead of classical AHP to tackle uncertainties. Classical AHP is a stochastic process, and fuzzy AHP, on the other hand, neither yields better result nor reduces the complexity of analysis. Several researchers including Saaty showed that fuzzy AHP has certain limitations. In this book, fuzzy version of different integrated mathematical models for supplier selection are discussed and the conflict of choice to opt classical AHP or fuzzy AHP is left to the reader. Rosenbloom (1996) further enhanced the capability of AHP with the probabilistic interpretation of final rankings. Usually, designers and manufacturers prefer to use interval range instead of exact dimension to mention size of product to cope with inevitable inaccuracy of manufacturing process. The concept of tolerance in metric space can also be used in decision space. Aguarón et al. (2003) adopted similar approach to propose **consistency stability interval (CSI)** to associate an interval range instead of a crisp value with every judgment of pairwise comparison. Causes for the validation of model can be divided into two categories:

1. *Internal cause*: They are the inherent part of every MCDA problem and some of them are as follows:
    1. Misinterpretation of actual problem brings ambiguity, vagueness, etc.
    2. Wrong selection of appropriate method yields erroneous result.
    3. Misleading judgment may accept the worst alternative instead of the best alternative.
2. *External cause*: They are external to the system and some of them are as follows:
    1. Wrong data collection method.
    2. Wrong interpretation of interdependencies among criteria and/or alternatives.

Therefore, decision-makers must find controllable and uncontrollable causes to give optimum result.

1.4.6 Different Forms of AHP

Extensive literature shows that researchers are more interested in using **hybrid AHP** instead of classical AHP to combine the benefits of more than one method. In Table 1.1, list of hybrid AHP is shown.

Table 1.1 Integrated AHP

Sl. no.	Author/s	Journal/conference name and year	Remarks
1	Ali Najmi and Ahmad Makui	International Journal of Industrial Engineering Computations (2010)	They proposed the integrated use of AHP and DEMATEL to tackle interdependencies of criteria
2	David P. Lilly, John Cory, Bill Hissem	Oxford Business and Economics Conference Program (2009)	Principal component analysis was developed by Pearson in 1901 and has been successfully adapted to many industries to create summary information. The principal component analysis method creates an overall measure, which is a refined dashboard type of performance measurement using statistical data compression techniques. The variable reduction technique creates artificial variables and reduces redundancy in the correlations
3	A.A. Zadeh, H.R. Izadbaksh	International Journal of Industrial Engineering (2008)	They proposed an integrated approach of PCA and AHP and further compared the performance of their proposed method with the integrated approach of DEA and AHP. As per their observation, the performance of DEA and AHP is much inferior to their proposed method in the context of plant layout
4	Zixue Guo and Yi Zhang	IEEE conference (2010)	They used integrated approach of AHP and PCA
5	Weijun Xia and Zhiming Wu	Omega (2007)	They used integrated approach of rough sets theory and AHP to enhance consistency
6	Ehsan Akhlaghi	World Academy of Science, Engineering and Technology (2011)	Used fuzzy RST
7	Chen-Guang Guo, Yong-Xian Liu,	International Journal of Automation and	Fuzzy AHP and RST

(continued)

Table 1.1 (continued)

Sl. no.	Author/s	Journal/conference name and year	Remarks
	Shou-Ming Hou and Wei Wang	Computing. November (2010)	
8	Te-Sheng Li and Hsing-Hsin Huang	Expert Systems with Applications (2009)	They proposed combined approach of fuzzy AHP and TRIZ for product design. In 1946, Altshuller, a mechanical engineer of Russian Navy, developed TRIZ, an acronym for the theory of inventive problem solving

1.4.7 Application of AHP

From early 1970s, AHP has been extensively used because of its computational simplicity, flexibility to be integrated with other MCDA tools, and strong mathematical background. In this regard, three review works since 1979 onward are identified on the application of AHP, as shown in Table 1.2.

Table 1.2 Application of AHP

Authors	Period	No. of papers referred	No. of application areas identified
Jung. P. Shim	1979–1988	141	31 different areas of application
Omkarprasad S. Vaidya and Sushil Kumar	Prior to 1990–2003	154	10 different areas are identified, and each area is further sub-divided into 9 sub-areas
Nachiappan Subramanian and Ramakrishnan Ramanathan	1990–2009	291	Area related to operations management are identified as 1. Operations strategy—manufacturing, technological, socioeconomical, and environmental strategies 2. Product and process design—product planning, forecasting, quality management, measuring and improving performance, layout of facilities, managing capacity 3. Planning and scheduling resources—resource allocation, job design, and work measurement 4. Project management—project estimation, project planning, project control 5. Managing the supply chain—logistics and supply chain management, outsourcing, managing stocks

1.4.8 Analytic Network Process (ANP)

Usually, every decision is influenced by its surrounding stimuli. On the other hand, ANP consists of several clusters and they are connected with each other by their dependence. Cluster basically groups elements with common set of attributes. According to Saaty (2004), ANP integrates human intuition and judgment with reason. In Fig. 1.1, a comparison of AHP and ANP is shown. Saaty broadly classified dependencies as inner dependencies and outer dependencies. Inner dependencies may exist among elements of cluster and outer dependencies may exist between two clusters.

$$W_n = \begin{bmatrix} 0 & 0 & 0 & 0 \\ W_{21} & 0 & 0 & 0 \\ 0 & W_{32} & 0 & I \end{bmatrix} \quad \text{for hierarchical model.}$$

$$W_n = \begin{bmatrix} 0 & 0 & 0 & 0 \\ W_{21} & W_{22} & 0 & 0 \\ 0 & W_{32} & W_{33} & I \end{bmatrix} \quad \text{for network model.}$$

In AHP, W_{22} and W_{33} are equal to zero. The steps of ANP are briefly mentioned below:

Step 1: Define goal or objective of the problem.
Step 2: Identify criteria and sub-criteria.
Step 3: Determine the inner dependencies between factors by pairwise comparison using 1–9 scale to calculate priority vectors.
Step 4: Determine the inner as well as outer dependencies between sub-factors by pairwise comparison using 1–9 scale to calculate priority vectors. For each comparison, inconsistency index should be less than 0.1 to accept the decision.
Step 5: Form unweighted super matrix by using priority vectors. In ANP, a stochastic super matrix is formed through the series of matrix operations.

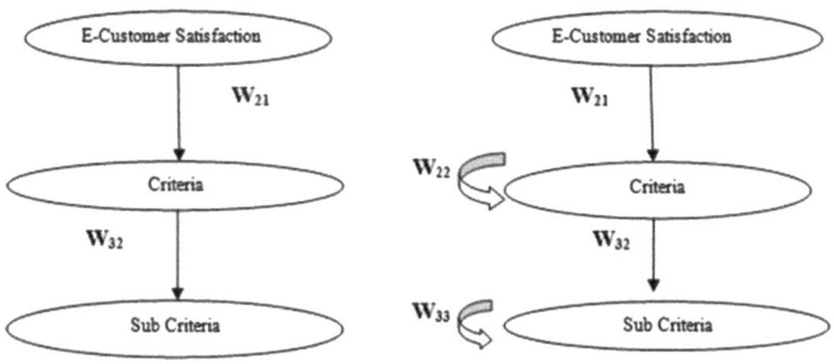

Fig. 1.1 Comparisons of AHP and ANP

Step 6: Form weighted super matrix by dividing each element by the sum of the corresponding column elements. So that the sum of each column will be equal to one.
Step 7: Form limit matrix by raising the power of weighted super matrix to an arbitrary high power. Eigenvalues in stochastic matrix are less than one (Saaty 2004). The sum of each column in limit matrix will be equal and the sum of limiting priorities will be equal to the number of criteria.
Step 8: Determine the normalized priority value of each sub-criteria by cluster.

Like AHP, rank reversal problem is also expected in ANP. Leung and Cao (2001) showed that **Sinarchy**, a particular form of analytic network process (ANP), could prevent rank reversal.

1.4.9 Technique for Order Preference by Similarity to Ideal Solution (TOPSIS)

TOPSIS, a three-point approach, is one of the most cited MCDM approaches to find solutions from the set of finite alternatives by minimizing the distance from an ideal point and maximizing the distance from a nadir point (Olson 2004; Opricovic and Tzeng 2004). However, relative importance of these distances is neglected in TOPSIS (Opricovic and Tzeng 2004). In 1981, Hwang and Yoon developed TOPSIS by modifying the method of Zeleny (1974). Hwang and Yoon (1981) proposed the following six steps to use TOPSIS:

1. Prepare of normalized decision matrix.
2. Prepe of weighted normalized decision matrix.
3. Identify positive ideal solution and negative ideal solution.
4. Determine the separation measure.
5. Determine relative closeness coefficient.
6. Rank the alternative.

Brain storming, **nominal group technique (NGT)**, **Delphi** technique, etc. can also be integrated with TOPSIS to enhance the quality of decision (Shih et al. 2007). One of the most important advantages of TOPSIS is its ability to identify the best alternative quickly (Parkan and Wu 1997). Like other MCDA tools, method of normalization for TOPSIS can be simplified as linear transformation (Saghafian and Hejazi 2005; Chen 2000). Extensive literature shows the following two different ways to use TOPSIS in fuzzy environment:

1. Defuzzification of ratings and weights.
2. Generalized fuzzy TOPSIS.

Usually, later method is considered better as it prevents the loss of information during defuzzification. Wang and Lee (2007) extended TOPSIS in fuzzy environment with two parameters, **Up and Lo**. Interval-valued fuzzy set theory, on the other hand, can provide a more accurate modeling. In this regard, researchers (Chen and Tsao 2008; Ashtiani et al. 2009; Chu and Lin 2009) used the extension of fuzzy TOPSIS method with interval-valued fuzzy sets. Wang and Lee (2009) further modified TOPSIS with an innovative weighting mechanism to avoid the subjectivity of DM's personal bias. Nezhad and Damghani (2009) used preference ratio to rank alternatives based on closeness coefficient in TOPSIS. They recommended that consistency rate between two distance measures gradually reduces with the increase of alternatives. Shih et al. (2007) used internal aggregation for TOPSIS. Taleizadeh et al. (2009) integrated fuzzy set theory, TOPSIS, and GA to solve inventory problem. Tsou (2008) used multi-objective particle swarm optimization (MOPSO) with TOPSIS to solve inventory problems. Lin et al. (2008) combined AHP and TOPSIS to map customer requirements with design characteristics of product. Shih et al. (2007) further compared the performance of TOPSIS and AHP in the presence of large number of criteria and confirmed that AHP can accommodate only 7 or 9 criteria but TOPSIS can accommodate many more. Several normalization processes, such as **vector normalization**, **linear normalization** and **non-monotonic normalization**, can also be used for TOPSIS along with two distance measures function—**Minkowski's Lp metrics** and **weighted Lp metrics** (Shih et al. 2007).

1.4.10 Fuzzy Hierarchical TOPSIS

In brief, fuzzy hierarchical TOPSIS has the following steps (Wang et al. 2008):

1. Fuzzy analytic hierarchy process uses a top-down structure to determine the fuzzy weight of each criterion.
2. TOPSIS prepares a normalized fuzzy performance matrix by using characteristic of each criterion and finally forms a normalized weighted performance matrix.
3. Calculate FPIS and FNIS, and apply the metric distance method to calculate the dispersion between alternatives under each criterion.
4. Finally, get a best ranking through an appropriate method of aggregation.

Fuzzy hierarchical TOPSIS can be briefly summarized as follows:

Step 1: Define the evaluation criteria and alternatives of the decision-making problem to frame a top-down structure.
Step 2: Prepare fuzzy pairwise comparison matrix for each criterion to evaluate alternatives.
Step 3: Use the Lambda-Max method to calculate the fuzzy weight (FAHP) of each criterion given by the experts.
Step 4: Check the consistency index (C.I.)
Step 5: Take geometric mean to aggregate all expert opinion.
Step 6: Normalize fuzzy performance matrix.
Step 7: Prepare weighted normalized fuzzy performance matrix.
Step 8: Calculate FPIS and FNIS.

Step 9: Measure the distance of each point from FPIS and FNIS.
Step 10: Apply the Euclidean distance method to aggregate all of the criteria for each alternative.
Step 11: Select the best alternative.

1.4.11 Rank Reversal Problem in TOPSIS

Like AHP, TOPSIS does have rank reversal problems. Cascales and Lamata (2012) identified the following reasons of rank reversal problem in TOPSIS:

1. Selection of norm.
2. Selection of positive ideal solution (PIS) and negative ideal solution (NIS).

Vector normalization, commonly used in classical TOPSIS, is represented as follows:

$$r_{ij} = \frac{x_{ij}}{\sum_{j=1}^{m} (x_{ij})^2} \quad \forall i = 1, 2, 3, \ldots, m \,\text{and}\, j = 1, 2, 3, \ldots, n$$

Chakraborty and Yeh (2009) mentioned that vector normalization is the most appropriate to maintain rank consistency and weight sensitivity, respectively. However, Cascales and Lamata (2012) proposed the modification of the vector normalization with new r_{ij} for preserving rank, where $r_{ij} = \frac{x_{ij}}{\max_i (x_{ij})} \forall i = 1, 2, \ldots, m \,\text{and}\, r_{ij} \leq 1$. They further mentioned that modification of norm and selection method for PIS and NIS is must for preserving rank. For example, after the addition of new alternative, if the normalized matrix $A = [r_{ij}]_{m \times n}$, where $\max(r_{ij}) = 1 \forall i = 1, 2, \ldots, m \,\text{and}\, j = 1, 2, \ldots, n$, then **positive ideal solution (PIS)** becomes $A^+ = [1, 1, \ldots, 1]$. Here, PIS remains unchanged. However, there are chances that negative ideal solution (NIS) $A^- = [\min(r_{ij})]$ may change. With the change in NIS, closeness coefficient as well as rank of alternatives changes.

1.4.12 TOPSIS and Other Methods

Today, researchers are more keen to use integrated or hybrid TOPSIS (Behzadian et al. 2012). Govindan et al. (2012) identified the following advantages of TOPSIS:

1. TOPSIS can evaluate large number of alternatives w.r.t large number of criteria.
2. Changes in one attribute can be compensated by other attributes as TOPSIS allows explicit balancing and interactions among attributes.

3. Instead of giving only rank of each alternative like other MADM tools, TOPSIS gives preferential ranking of alternatives with a numerical value to give better understanding of differences and similarities between alternatives.
4. According to the simulation comparison from Zanakis et al. (1998), TOPSIS has the fewest rank reversals among all MADM tools.

To identify the trend of research on TOPSIS, work of Behzadian et al. (2012) is further extended to show the distribution of research papers on hybrid methods of TOPSIS since 2000 in Fig. 1.2.

As shown in Fig. 1.2, about 33.33% of total research work has been allotted to the integrated use of fuzzy set theory and TOPSIS and about 14.86% of the total research work on hybrid TOPSIS has been allotted to the integrated approach of AHP and TOPSIS.

1.4.13 Application of TOPSIS

Behzadian et al. (2012) identified the following application areas of TOPSIS:

1. Supply chain management and logistics
2. Design, engineering, and manufacturing systems
3. Business and marketing management
4. Health, safety, and environment management
5. Human resources management
6. Energy management

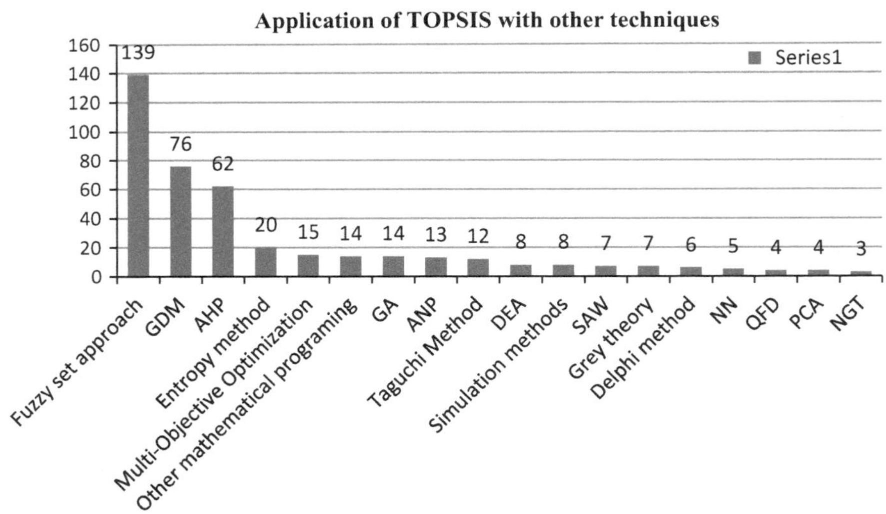

Fig. 1.2 Distribution of research papers on hybrid TOPSIS since 2000 onward

1.4.14 VIKOR (VlseKriterijumska Optimizacija I Kompromisno Resenje; in Serbian)

VlseKriterijumska Optimizacija I Kompromisno Resenje or VIKOR means multi-criteria optimization and compromise solution (Chu et al. 2007). It ranks alternatives w.r.t 'closeness' to the 'ideal' solution (Opricovic 1998, as stated in Oprocovic and Tzeng 2004). **VIKOR** is helpful in situation where decision-makers do not know their preferences at the beginning of the system (Opricovic and Tzeng 2004). In VIKOR, '**compromise**' means a feasible solution space closest to ideal solution and compromise means an agreement established by mutual consensus (Opricovic and Tzeng 2007, as stated in Liu and Wang 2011). There are several methods that have been developed since 1998 such as

1. Fuzzy VIKOR
2. Extended VIKOR
3. OWA (ordered weighted averaging aggregation)—VIKOR
4. Other MCDA tools are integrated with VIKOR.

Opricovic and Tzeng (2004) compared TOPSIS and VIKOR based on four criteria—procedural basis, method of normalization, aggregation, and solution. VIKOR is one of the compromise programming which has received much attention from different fields, namely manufacturing, design, finance, and health care.

1.5 Uncertainty Analysis with MCDA

Uncertainty refers to the absence of precise information. Often, vagueness and imprecision are used to refer uncertainty in measurement. Vagueness and imprecision differ in representing the form of uncertainty. Vagueness usually relates to non-measureable or qualitative data. For example, the quality of supply is good. The term 'good' is a linguistic variable to refer the level of satisfaction associated with the product. For example, the delivery date of supply is 10–15 days. For a planned horizon, if we plot the delivery date of supply, then we can state the imprecision in probabilistic terms. Uncertainty can be dealt with several theories such as **fuzzy set**, **vague set**, **gray set**, **intuitionistic fuzzy set**, **probability theory**, **possibility theory**, and **evidence theory**. Each of them is used to deal with different types of uncertainty. Klir and Yuan identified three different types of uncertainty—**fuzziness**, **discord**, and **non-specificity**. Fuzzy set represents fuzziness, probability distribution represents only discord, and classical set represents non-specificity (Jousselme et al. 2005). Classical MCDA tool uses crisp set during pairwise comparison of criteria, sub-criteria and alternative with respect to each sub-criteria/criteria. Uncertainty is associated with assigning a crisp value during pairwise comparison. Moreover, in real-life problem, every data, be cardinal or ordinal, is associated with different types of uncertainty. Every human being does make decision under certainty, uncertainty, and risk (Taha 2006, p. 503), and in each situation, different methods are used, as shown in Table 1.3.

In decision making, uncertainty arises when the consequences of an action are unknown as it depends on future events. Durbach and Stewart (2012) classified uncertainty into two categories—**internal uncertainty** and **external uncertainty**. Internal uncertainty refers to uncertainty about decision-maker's preferences, problem identification, vagueness, imprecise information, etc. External uncertainty refers to uncontrollable events of surrounding which could affect the outcome of any decision.

1.5.1 Fuzzy Set—An Introduction

Concept of set is the crux of mathematics. It is difficult to define set as in mathematics it is not well defined yet. Set can be defined as collection of well-defined objects. A number a is the member of the set A or not is defined by its *characteristic function* $U_A(a)$ which is mentioned below:

$$U_A(a) = \begin{cases} 1 & \text{where } a \in A \\ 0 & \text{where } a \notin A \end{cases}$$

Zadeh proposed that there are infinite possible values between the real continuous intervals [0,1] and introduced the concept of 'degree of membership.' The sets on the universe U that encompasses '**degree of membership**' are called as fuzzy set. If $u \in \boldsymbol{U}$ and $A(u)$ is degree of membership then fuzzy set $\tilde{A} = \{u, A(u)\}$. The selection of membership functions is highly subjective and sensitive to the application area. Let us consider the price of a product is close to Rs. 6 and it varies from Rs. 5 to Rs. 7. An expert can select either of the two membership functions to express above statement, as shown in Fig. 1.3.

Table 1.3 Types of decision

Types of decision	Types of payoff matrix	Method
Decision making under certainty	All elements of payoff matrix are well defined	Linear programming
Decision making under risk	Payoff associated with each decision alternatives is described by probability distribution	1. **Decision tree analysis** 2. Utility functions
Decision making under uncertainty	All elements of payoff matrix are random in nature	**Laplace** **Minimax** **Savage** **Hurwicz**

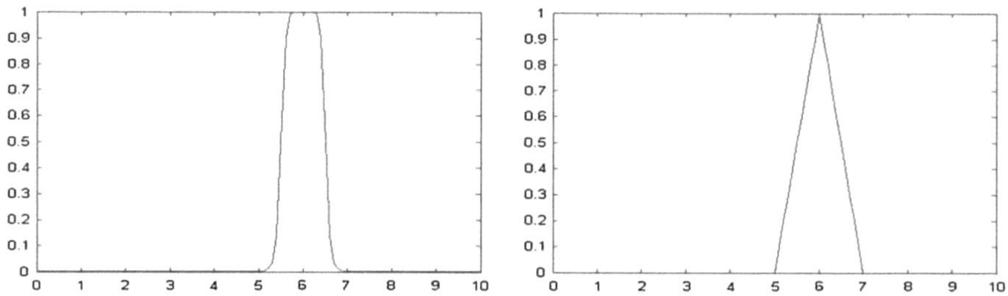

Fig. 1.3 Examples of membership functions

Usually, triangular fuzzy shape is considered for its simplicity and stability during defuzzification process. A positive TFN $\tilde{n}$ can be specified by three parameters (a,b,c), as shown in Fig. 1.4. The membership function $\mu_{\tilde{A}}$ is defined as

$$\mu_A = \begin{cases} 0, & x \leq a \\ \frac{x-a}{b-a}, & a \leq x \leq b \\ \frac{c-x}{c-b}, & b \leq x \leq c \\ 0, & c \leq x \end{cases}$$

If $\tilde{a} = (a_1, a_2, a_3)$ and $\tilde{b} = (b_1, b_2, b_3)$ are two triangular fuzzy numbers (TFNs), then the distance between them calculated by vertex method is as follows:

$$d(\tilde{o}, \tilde{n}) = \sqrt{\frac{1}{3}\left[(a_1 - b_1)^2 + (a_2 - b_2)^2 + (a_3 - b_3)^2\right]}$$

Actually, defuzzified value varies for all fuzzy membership functions for different methods except TFN, as shown in Table 1.4.

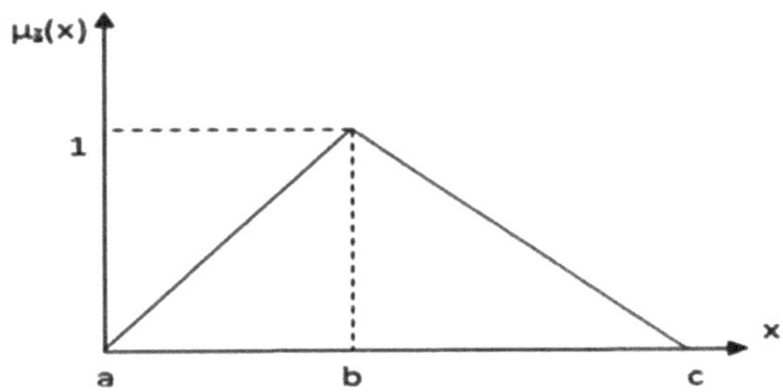

Fig. 1.4 Triangular fuzzy number

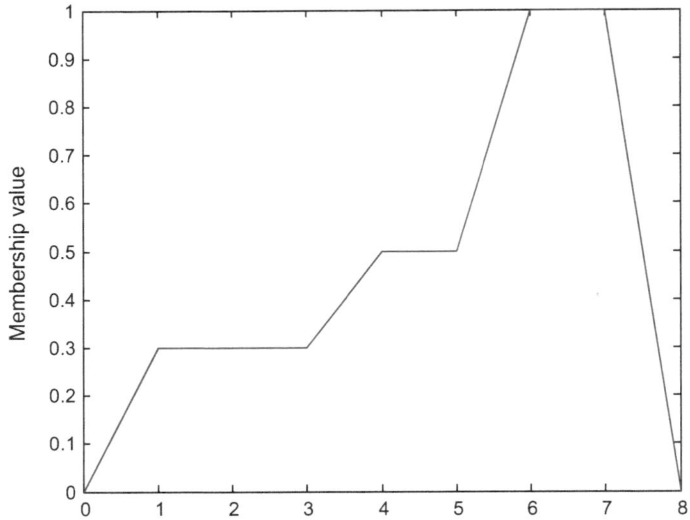

Fig. 1.5 Custom membership function

1.5.2 Cascaded Fuzzy Inference System

Knowingly or unknowingly, human being prefers to use linguistic variable to express his shallow or imprecise knowledge about any fact or objects such as *about to*, *very small*, and *fairly*. With this natural expression, we conclude or make decision in uncertain environment. In fuzzy inference system (FIS), such expressions are formed with If-then rule as mentioned below:

If premise, then conclusion or if x is $\tilde{A}_1$ and $\tilde{A}_2$ Then y is $\tilde{B}$

The above rule can be expressed as if x is $\tilde{A}$ Then y is $\tilde{B}$

where $\tilde{A} = \widetilde{A_1} \cap \widetilde{A_2} = \min[\mu_{\tilde{A_1}}(x), \mu_{\tilde{A_2}}(x)]$

The above premise or antecedent is known as **multiple conjunctive antecedent**. Similarly, if x is $\tilde{A}$ Then y is $\tilde{B}$, where $\tilde{A} = \widetilde{A_1} \cup \widetilde{A_2} = \max[\mu_{\tilde{A_1}}(x), \mu_{\tilde{A_2}}(x)]$

The above antecedent is known as **multiple disjunctive antecedent**. Fuzzy inference system can be broadly classified as follows: (1) **Mamdani systems**, (2) **Sugeno models**, and (3) **Tsukatomoto models**. Mamdani models can be further classified depending upon the method of aggregation, defuzzification, etc. Depending upon the types of output function, Sugeno models can be classified as **zero-order Sugeno**, **first-order Sugeno,** etc. For several reasons, Mamdani models are preferred. In this section, a fuzzy inference system is developed with two-input and one-output Mamdani model. One of the main drawbacks of fuzzy inference system using **intersection rule configuration (IRC)** is the combinatorial explosion of rules (Ross 2007, p. 75). If N is the number of rules, K is the number of linguistic variables, and n is the number of input variables, then $N = K^n$. Thus, the explosion

Table 1.4 Comparison of different defuzzification methods

Defuzzification methods						
Types of fuzzy member	Centroid	Bisector	Mean of maximum (MOM)	Smallest of maximum (SOM)	Largest of maximum (LOM)	Remark
TFN (1,3,5)	3	3	3	3	3	For TFN, defuzzified value remains the same for all methods. Time taken in defuzzification is as follows: LOM < SOM < CENTROID < MOM < BISECTOR
Time in seconds	0.000432	0.003799	0.002213	0.00243	0.00023	
TRAPMF (1,2,3,4)	2.5	2	2.5	2	3	For trapezoidal membership functions, defuzzified values varies. Time taken in defuzzification is as follows: BISECTOR < LOM < SOM < CENTROID < MOM
Time in seconds	0.0005	0.000017	0.0000584	0.000052	0.000041	
Custom functions	4.9847	6	6.5	6	7	For custom function shown in Fig. 1.5 gives different defuzzified values for different methods. Time taken in defuzzification is as follows: LOM < SOM < CENTROID < MOM < BISECTOR

of rules can be avoided by reducing K and n. In three different ways, combinatorial rule explosion can be avoided (Davidrajuh 2008):

1. Dividing the fuzzy inference system into multiple numbers of sub-systems. Instead of preparing a monolithic FIS with large number of rules, it is advisable to split into several layers. This approach is known as cascaded fuzzy inference system.
2. Use of limited number of rules.
3. Use of limited number of inputs.

In cascaded fuzzy inference system, number of rules in each layer or stage can be calculated with the formula mentioned below (Davidrajuh 2008):

$$\left(m^{N/n_1} \times n_1\right) + \left(m^{n_1/n_2} \times n_2\right) + \cdots + \left(m^{n_{r-1}/n_r} \times n_r\right) \tag{1.1}$$

where N is the number of input, n_i is the number of sub-system in ith layer, r is the number of layer of sub-systems, and m is the number of fuzzy membership functions for each input. Still the performance of cascaded fuzzy inference system can be improved by reducing the number of rules in **fuzzy associative memory (FAM)** by using either **singular value decomposition (SVD)** or **Combs** method as **cascaded fuzzy inference system** is the collection of several fuzzy inference systems in hierarchical form. In cascaded fuzzy inference system, the output from one FIS is fed to other FIS after processing. Thus, the explosion of rule base is quite expected if it is developed improperly.

1.5.3 Intuitionistic Fuzzy Set (IFS)—An Introduction

In 1986, **Atanassov** proposed **intuitionistic fuzzy set (IFS),** the generalized concept of fuzzy set. If U be a universe of discourse, then IFS A can be defined as $A = \{(u, \mu_A(u), \nu_A(u)) | u \in U\}$, where $\mu_A(u)$ and $\nu_A(u)$ denote membership and non-membership functions of A, respectively, and satisfy $0 \leq \mu_A(u) + \nu_A(u) \leq 1$ $\forall u \in U$. For every IFS A in U, **degree of hesitation** can be defined as $\pi_A(u) = 1 - \mu_A(u) - \nu_A(u)$ which expresses whether u belongs to A or not.

Any element u that belongs to IFS A should be inside the triangle MNO, as shown in Fig. 1.6.

If $A = \{(u, \mu_A(u), \nu_A(u)) | u \in U\}$ and $B = \{(u, \mu_B(u), \nu_B(u)) | u \in U\}$, then normalized hamming distance between A and B can be represented as $l(A, B) = \frac{1}{2n} \sum_{i=1}^{n} [|\mu_A(u) - \mu_B(u)| + |\nu_A(u) - \nu_B(u)| + |\pi_A(u) - \pi_B(u)|]$ (Szmidt and Kacprzyk 2000). To rank three IFS, their normalized hamming distance from the ideal solution M (1,0,0) should be calculated. Lowest distance from M will give better solution. Every human decision is associated with a degree of hesitation. If degree of hesitation is zero, then it is certainty. Essence of intuitionistic fuzzy set is to measure this degree of hesitation associated with human decision. Szmidt and Kacprzyk (2000) mentioned

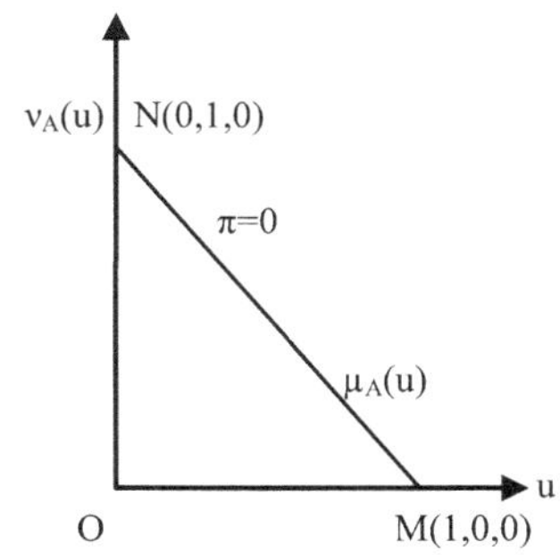

Fig. 1.6 2-D representation of IFS

graphical comparison of fuzzy set and intuitionistic fuzzy set, and similar comparison was prepared with MATLAB, as shown in Fig. 1.7, to explain generalized IFS.

As shown in Fig. 1.7, an intuitionistic fuzzy set U is mapped into the triangle ABC, shaded in green color, so that each element of U is corresponding to an element of ABC. Point P refers to all points belonging to intuitionistic fuzzy set, and point A refers to all points not belonging to intuitionistic fuzzy set. If orthogonal projection of ABC is taken on plane, triangle A°B°C°, shaded in blue color, is obtained.

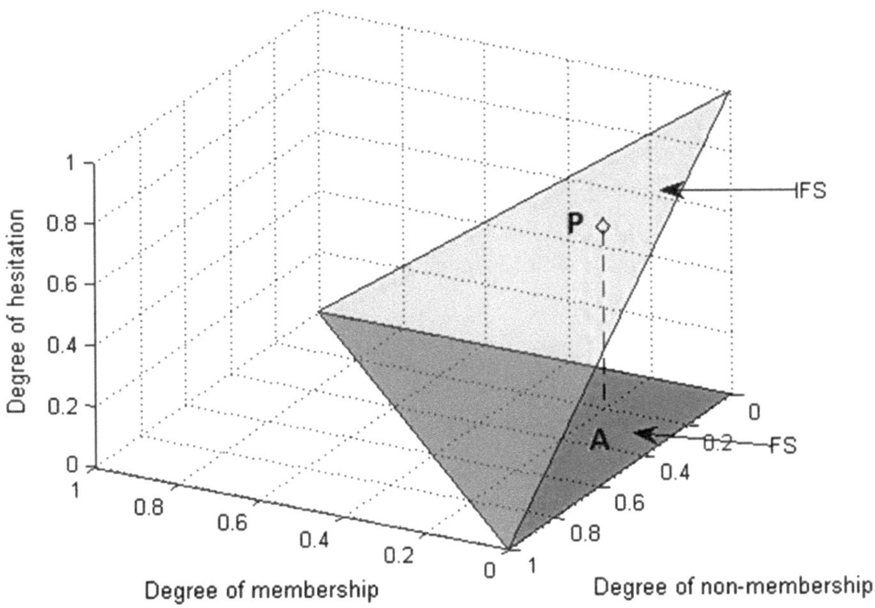

Fig. 1.7 Graphical presentation of fuzzy set and intuitionistic fuzzy set (Color Online)

1.5.4 Dealing Uncertainty with AHP

There are several forms of AHP to deal with uncertainties. Some of them are mentioned below:

1. Extent fuzzy AHP
2. Modified extent fuzzy AHP
3. Intuitionistic fuzzy AHP (IF-AHP)
4. Fuzzification of AHP by Shannon entropy
5. Gray AHP
6. Rough set theory and AHP

1.5.5 Dealing Uncertainty with TOPSIS

Like AHP, different approaches of TOPSIS are available to deal with uncertainty. Some of them are as follows:

1. Fuzzy hierarchical TOPSIS
2. Extent fuzzy TOPSIS
3. Fuzzy TOPSIS approach with Shannon entropy

1.5.6 Dealing Uncertainty with VIKOR

Like AHP and TOPSIS, the following approaches of VIKOR are available to deal with uncertainties. Some of them are as follows:

1. Extent fuzzy VIKOR
2. Fuzzy VIKOR approach with Shannon entropy
3. Gray VIKOR

1.5.7 Fuzzy AHP by Hand Calculation

Let us consider a simple three-criteria (C_1, C_2, and C_3) and three-alternative (A_1, A_2, and A_3) problem. Fuzzy linguistic values used for pairwise comparison are shown in the table.

Linguistic variables (from Tolga et al. 2005; with kind permission from Elsevier Limited).

Linguistic values	Triangular fuzzy numbers (*l*,*m*,*u*)
Equal	(1,1,1)
Weak	(2/3,1,3/2)
Fairly strong	(3/2,2,5/2)
Very strong	(5/2,3,7/2)
Absolute	(7/2,4,9/2)

Fuzzy pairwise comparison matrix for criteria is as follows:

Criteria	C_1	C_2	C_3
C_1	Equal (1,1,1)	Very strong (5/2,3,7/2)	Absolute (7/2,4,9/2)
C_2	——	Equal (1,1,1)	Fairly strong (3/2,2,5/2)
C_3	——	——	Equal (1,1,1)

Complete fuzzy pairwise comparison matrix is as follows:

Criteria	C_1	C_2	C_3
C_1	(1,1,1)	(5/2,3,7/2)	(7/2,4,9/2)
C_2	(2/7,1/3,2/5)	(1,1,1)	(3/2,2,5/2)
C_3	(2/9,1/4,2/7)	(2/5,1/2,2/3)	(1,1,1)

A triangular fuzzy number (TFN) contains three components—l,m,u. The above fuzzy pairwise comparison matrix prepared with the following relations is as follows:

$$\text{for every } i \neq j \begin{cases} l_{ij} = \frac{1}{u_{ji}} \\ m_{ij} = \frac{1}{m_{ji}} \\ u_{ij} = \frac{1}{l_{ji}} \end{cases}$$

Fuzzy synthetic extent value

S_{C1}	$(7,8,9) \otimes (1/14.85, 1/13.08, 1/11.41)$	(0.471,0.61,0.789)
S_{C2}	$(2.79,3.33,3.9) \otimes (1/14.85, 1/13.08, 1/11.41)$	(0.188,0.255,0.342)
S_{C3}	$(1.62,1.75,1.95) \otimes (1/14.85, 1/13.08, 1/11.41)$	(0.109,0.134,0.171)

Degree of possibility

$V(S_{C1} > S_{C2})$	1	min{1,1} = 1	Normalized priority of three criteria $W = [1,0,0]^T$
$V(S_{C1} > S_{C3})$	1		
$V(S_{C2} > S_{C1})$	0	min{0,1} = 0	
$V(S_{C2} > S_{C3})$	1		
$V(S_{C3} > S_{C1})$	0	min{0,0} = 0	
$V(S_{C3} > S_{C2})$	0		

Similarly, normalized priority of alternatives can be calculated as shown above. The final normalized priority for three alternatives is 0.57, 0.16, and 0.27.

1.6 Conclusion

In this chapter, a brief introduction is given to MCDA. Well-known MCDA tools such as AHP, ANP, TOPSIS, VIKOR, and their fuzzy versions are discussed in detail. A detail discussion is given on rank reversal problem of TOPSIS, various applications of AHP, ANP, TOPSIS, and VIKOR, uncertainty analysis with AHP, TOPSIS, VIKOR, etc.

Recent trend of research related to application and development of MCDA tools shows that researchers are more interested for the integrated application of MCDA tools. However, such application raises questions about the legitimacy of integrated approach, particularly for the combined application of fuzzy set theory and AHP. Still integrated model of fuzzy AHP-TOPSIS, fuzzy AHP-VIKOR, and IF-AHP is discussed with illustrative case study, and choice of selection of appropriate method is left to the interested readers. Limitations of extent fuzzy AHP are discussed at the end of Chap. 4. A brief discussion is also included in this chapter.

Few researchers also showed preference to use cascaded fuzzy inference system to select suppliers in complex decision-making process to avoid subjectivity of human assessments during pairwise comparison. Such subjectivity of human assessments during pairwise comparison can be avoided by considering a range of value instead of a crisp value. In this regard, author would like to say that the justification of usefulness of any MCDA tool is a priori of decision-making process. In Sect. 1.5.1, a brief introduction to fuzzy set theory is given with its certain limitations. Selection of fuzzy membership functions is still subjective. Triangular fuzzy membership functions are commonly used for the simplicity of calculation and stability of defuzzification. Fuzzy set theory is well accepted in different engineering applications but its application in decision sciences is not unanimously accepted yet. For example, AHP is a stochastic process. Fuzzy AHP brings more fuzziness to result and thereby computational complexity with erroneous result. In this book, several methods are discussed and selection of appropriate method is left to the reader.

The following useful findings are mentioned in this chapter:

- Steps of decision-making process, classification of criteria, etc. are explained with examples to give brief introduction to the readers.
- Different scales, normalization methods, etc. are mentioned in this chapter.
- Rank reversal problem is discussed in detail for AHP, ANP, and TOPSIS as it is one of the debated issues of MCDA.
- Recent trend of MCDA is discussed for AHP, ANP, TOPSIS, and VIKOR.
- Three hybrid algorithms are discussed for AHP, TOPSIS, and VIKOR.
- Several comparative analyses are shown in tabular format in this chapter such as AHP-TOPSIS and TOPSIS-VIKOR.
- Introduction to cascaded fuzzy inference system is included and the same is discussed in detail Chap. 4 with an illustrative example.
- Finally, a brief discussion is given to deal with uncertainty with MCDA tools such as AHP, TOPSIS, and VIKOR.

References

Aguarón J, Esocbar MT, Jiménez JMM (2003) Consistency stability intervals for a judgement in AHP decision support systems. Eur J Oper Res 145:382–393

Akhlaghi E (2011) A rough-set based approach to design an expert system for personnel selection. World Academy of Science, Engineering and Technology 78

Ashtiani B, Haghighirad F, Makui A, Montazer GA (2009) Extension of fuzzy TOPSIS method based on interval-valued fuzzy sets. Appl Soft Comput 9:457–461

Baker D, Bridges D, Hunter R, Johnson G, Krupa J, Murphy J, Sorenson K (2002) Guidebook to decision-making methods. WSRC-IM-2002-00002, Department of Energy, USA. http://emi-web.inel.gov/Nissmg/Guidebook_2002.pdf

Barzilai J, Lootsma FA (1997) Power relations and group aggregation in the multiplicative AHP and SMART. J Multi-Criteria Decis Anal 6:155–165

Behzadian M, Otaghsara SK, Yazdani M, Ignatius J (2012) A state-of the-art survey of TOPSIS applications. Expert Syst Appl 39:13051–13069

Belton V, Gear T (1982) On a shortcoming of Saaty's method of analytic hierarchies. Omega 11 (3):226–230

Beynon M (2002) An analysis of distributions of priority values from alternative comparison scales within AHP. Eur J Oper Res 140:104–117

Cascales MSG, Lamata MT (2012) On rank reversal and TOPSIS method. Math Comput Model 56:123–132

Chakraborty S, Yeh CH (2009) A simulation comparison of normalization procedures for TOPSIS. IEEE. ISSN: 978-1-4244-4136-5/09

Chen CT (2000) Extensions of the TOPSIS for group decision making under fuzzy environment. Fuzzy Sets Syst 114:1–9

Chen TY, Tsao CY (2008) The interval-valued fuzzy TOPSIS method and experimental analysis. Fuzzy Sets Syst 159:1410–1428

Choo EU, Wedley WC (2004) A common framework for deriving preference values from pair wise comparison matrices. Comput Oper Res 31:893–908

Chu TC, Lin YC (2009) An interval arithmetic based fuzzy TOPSIS model. Expert Syst Appl 36:10870–10876
Chu MT, Shyu J, Tzeng GH, Khosla R (2007) Comparison among three analytical methods for knowledge communities group-decision analysis. Expert Syst Appl 33(4):1011–1024
Davidrajuh R (2008) Building a Fuzzy Logic based Tool for E-readiness measurement. Electron Gov Int J 5(1):120–130
De Luca A, Termini S (1972) A definition of a non-probabilistic entropy in the setting of fuzzy sets theory. Inf Control 20:201–312
Dodd F, Donegan H (1995) Comparison of prioritization techniques using inter hierarchy mappings. J Oper Res Soc 46:492–498
Dong Y, Xu Y, Li H, Dai M (2008) A comparative study of the numerical scales and the prioritization methods in AHP. Eur J Oper Res 186:229–242
Durbach IN, Stewart TJ (2012) Modeling uncertainty in multi-criteria decision analysis. Eur J Oper Res 223:1–14
Govindan K, Khodaverdi R, Jafarian A (2012) A fuzzy multi criteria approach for measuring sustainability performance of a supplier based on triple bottom line approach. J Clean Prod. doi:10.1016/j.jclepro.2012.04.014
Guitouni A, Martel JM (1998) Tentative guidelines to help choosing an appropriate MCDA method. Eur J Oper Res 109:501–521
Guo Z, Zhang Y (2010) The third-party logistics performance evaluation based on the AHP-PCA model. IEEE. ISBN:978-1-4244-7161-4/10
Guo CG, Liu YX, Hou SM, Wang W (2010) Innovative product design based on customer requirement weight calculation model. Int J Autom Comput 7(4):578–583. doi:10.1007/s11633-010-0543-3
Harker P, Vargas L (1987) The theory of ratio scale estimation: Saaty's analytic hierarchy process. Manage Sci 33:1383–1403
Hwang CL, Yoon K (1981) Multiple attribute decision making. Springer-Verlag, Berlin
Ishizaka A, Labib A (2011) Review of the main developments in the analytic hierarchy process. Expert Syst Appl 38:14336–14345
Ishizaka A, Balkenborg D, Kaplan T (2010) Influence of aggregation and measurement scale on ranking a compromise alternative in AHP. J Oper Res Soc 62:700–710
May EC, Spottiswoode SJP, James, CL (1994) Shannon entropy: a possible intrinsic target property. J Parapsychology 58
Jousselme A-L, Liu C, Grnecier D, Bossé Ẻ (2005) Measuring ambiguity in the evidence theory. IEEE Trans Syst Man Cybern Part A Syst Hum. doi:10.1109/TSMCA.2005.853483
Jr JT, Delhaye C, Kunsch PL (1989) An interactive decision support system (IDSS) for multicriteria decision aid. Math Comput Model 12(10/11):1311–1320
Leung LC, Cao D (2001) On the efficacy of modeling multi-attribute decision problems using AHP and Sinarchy. Eur J Oper Res 132:39–49
Li TS, Huang HH (2009) Applying TRIZ and Fuzzy AHP to develop innovative design for automated manufacturing systems. Expert Syst Appl 36:8302–8312
Lilly DP, Cory J, Hissem B (2009) The use of principal component analysis to integrate blasting into the mining process. In: Proceedings of 2009 Oxford Business & Economics Conference Program. ISBN: 978-09742114-1-9
Lin MC, Wang CC, Chen MS, Chang CA (2008) Using AHP and TOPSIS approaches in customer-driven product design process. Comput Ind 59:17–31
Liu P, Wang M (2011) An extended VIKOR method for multiple attribute group decision making based on generalized interval-valued trapezoidal fuzzy numbers. Sci Res Essays 6(4):766–776
Lootsma F (1989) Conflict resolution via pair-wise comparison of concessions. Eur J Oper Res 40:109–116
Ma D, Zheng X (1991) 9/9-9/1 Scale method of AHP. In: Proceedings of Second International Symposium on AHP, Pittsburgh
Najmi A, Makui A (2010) Providing hierarchical approach for measuring supply chain performance using AHP and DEMATEL methodologies. Int J Ind Eng Comput 1:199–212

Nezhad SS, Damghani KK (2009) Application of a fuzzy TOPSIS method base on modified preference ratio and fuzzy distance measurement in assessment of traffic police centers performance. Appl Soft Comput. doi:10.1016/j.asoc.2009.08.036

Olson DL (2004) Comparison of weights in TOPSIS Models. Math Comput Model 40:721–727

Opricovic S (1998) Multicriteria optimization of civil engineering systems. Faculty of Civil Engineering, Belgrade

Opricovic S, Tzeng GH (2004) Compromise solution by MCDM methods: a comparative analysis of VIKOR and TOPSIS. Eur J Oper Res 156:445–455

Opricovic S, Tzeng GH (2007) Extended VIKOR method in comparison with outranking methods. Eur J Oper Res 178(2):514–529

Parkan C, Wu ML (1997) On the equivalence of operational performance measurement and multiple attribute decision making. Int J Prod Res 35(11):2963–2988

Qureshi ME, Harrison SR, Wegener MK (1999) Validation of multicriteria analysis models. Agric Syst 62:105–116

Rosenbloom ES (1996) A probabilistic interpretation of the final rankings in AHP. Eur J Oper Res 96:371–378

Ross TJ (2007) Fuzzy logic with engineering applications. Wiley India Edition

Roy B (1985) Mèthodologie Multicritère D'Aide à la Dècision. Ćollection Gestion—Edition Economica, Paris

Saaty T (1977) A scaling method for priorities in hierarchical structures'. J Math Psychol 15 (3):234–281

Saaty TL (1990) Eigenvector and logarithmic least squares. Eur J Oper Res 48:156–160

Saaty TL (1994) Highlights and critical points in the theory and application of the analytic hierarchy process. Eur J Oper Res 74:426–447

Saaty TL (2004) Decision making—the analytic hierarchy and network process (AHP/ANP). J Syst Sci Syst Eng 13(1):1–34

Saaty TL, Shih HS (2009) Structures in decision making: on the subjective geometry of hierarchies and networks. Eur J Oper Res 199:867–872

Saaty TL, Vargas LG (1984) The legitimacy of rank reversal. OMEGA Int J Manag Sci 12(5):513–516

Saghafian S, Hejazi SR (2005) Multi-criteria group decision making using a modified fuzzy TOPSIS procedure. In: Proceeding 2005 International Conference on Computational Intelligence for Modelling, Control and Automation, and International Conference on Intelligent Agents, Web Technologies and Internet Commerce (CIMCA-IAWTIC'05). ISSN 0-7695-2504-0/05

Salo A, Hamalainen R (1997) On the measurement of preference in the analytic hierarchy process. J Multi-Criteria Decis Anal 6:309–319

Schoner B, Wedley WC (1989) Ambiguous criteria weights in AHP: consequences and solutions. Decis Sci 20:462–475

Schoner B, Choo EU, Wedley WC (1997) A comment on 'rank disagreement: a comparison of multi-criteria methodologies'. J Multi-Criteria Decis Anal 6:197–200

Shih HS, Shyur HJ, Lee ES (2007) An extension of TOPSIS for group decision making. Math Comput Model 45:801–813

Shim JP (1989) Bibliographical research on the analytic hierarchy process (AHP). Socio-Econ Plann Sci 23(3):161–167

Srdjevic B (2005) Combining different prioritization methods in the analytic hierarchy process synthesis. Comput Oper Res 32:1897–1919

Subramanian N, Ramanathan R (2012) A review of applications of analytic hierarchy process in operations management. Int J Prod Econ 138:215–241

Szmidt E, Kacprzyk J (2000) Distance between intuitionistic fuzzy sets. Fuzzy Sets System 114 (3):505–518

Taha HA (2006) Operations research—an introduction. Prentice-Hall of India Pvt, Ltd

Taleizadeh AA, Niaki STA, Aryanezhad MB (2009) A hybrid method of Pareto, TOPSIS and genetic algorithm to optimize multi-product multi-constraint inventory control systems with random fuzzy replenishments. Math Comput Model 49:1044–1057

Tolga E, Demircan ML, Kahraman C (2005) Operating system selection using fuzzy replacement analysis and analytic hierarchy process. Int J Prod Econ 97:89–117

Tsou CS (2008) Multi-objective inventory planning using MOPSO and TOPSIS. Expert Syst Appl 35:136–142

Vaidya OS, Kumar S (2006) Analytic hierarchy process: An overview of applications. Eur J Oper Res 169:1–29

Wang YM, Elhag TMS (2006) An approach to avoiding rank reversal in AHP. Decis Support Syst 42:1474–1480

Wang YJ, Lee HS (2007) Generalizing TOPSIS for fuzzy multiple-criteria group decision-making. Comput Math Appl 53:1762–1772

Wang TC, Lee HD (2009) Developing a fuzzy TOPSIS approach based on subjective weights and objective weights. Expert Syst Appl 36:8980–8985

Wang JW, Cheng CH, Cheng HK (2008) Fuzzy hierarchical TOPSIS for supplier selection. Appl Soft Comput. doi:10.1016/j.asoc.2008.04.014

Whitaker R (2007) Validation examples of the analytic hierarchy process and analytic network process. Math Comput Model 46:840–859

Xia W, Wu Z (2007) Supplier selection with multiple criteria in volume discount Environments. Omega 35:494–504

Yager RR (1979) On the measure of fuzziness and negation Part 1: membership in unit interval. Int J Gen Syst 5:21–229

Zadeh AA, Izadbaksh HR (2008) A multi-variate/ multi-attribute approach for plant layout design. Int J Ind Eng 15(2):143–154

Zanakis SH, Solomon A, Wishart N, Dublish S (1998) Multi-attribute decision making: a simulation comparison of selection methods. Eur J Oper Res 107:507–529

Zeleny M (1974) A concept of compromise solutions and the method of the displaced ideal. Comput Oper Res 1:479–496

Chapter 2
Modeling and Optimization of Traditional Supplier Selection

2.1 Introduction

Judicious selection of supplier can mitigate upstream supply chain risk by supplying right quantity at right place and time.

Supplier selection is a strategic process as it can mitigate upstream supply chain risk partially, if not completely. Better supplier–buyer dyadic relationship can enhance supply chain visibility and capability to cope with high demand volatility. Supplier selection, thus, is an indispensible part of any business. Any disruption in upstream supply may cause tremendous disaster in entire supply chain and compel organization to take risk. Risks in supply chain are broadly classified as **internal risk** that appears in normal operation and **external risk** that come from outside the supply. Selection of right supplier(s) could minimize external risks. Supplier selection could be either **single sourcing** or **multiple sourcing**. In single sourcing, entire supply comes from one supplier. In multiple sourcing, on the other hand, entire supply comes from a group of suppliers. Risk in supply chain could be minimized by internal integration and external integration of supply chain entities. External integration strongly encourages single sourcing by strengthening buyer–supplier relationship. Table 2.1 shows the comparative analysis of single-sourcing and multiple-sourcing strategies.

Research on supplier selection methods has rich collection, as shown in Fig. 2.1. Some researchers combined at least two techniques for supplier selection, for instance, AHP-GP, AHP-LP, DEA-AHP, and DEA-MOP. Supplier selection problem involves vague and imprecise assessments, which are by nature fuzzy. Thus, a group of researchers used fuzzy AHP. Various methods have been used to derive priority vectors from fuzzy pairwise comparison. A partial list is shown in Table 2.2.

Among all techniques, extent fuzzy AHP is used most frequently because of its computational simplicity. In Table 2.3, various techniques are classified based on single-sourcing and multiple-sourcing supplier selection.

K. Mukherjee, *Supplier Selection*, Studies in Systems, Decision and Control 88,
DOI 10.1007/978-81-322-3700-6_2

Table 2.1 Single sourcing versus multiple sourcing

Single sourcing	Multiple sourcing
Concept of this strategy comes from **just-in-time (JIT)** philosophy. Uncertainty in supply is very high as buyer deals with single supplier	Multiple sourcing is preferable if reliability of one supplier is very poor. It reduces safety stock without increasing stock-out problem (Kelle and silver 1990). It reduces uncertainty in supply but increases ordering cost (Agrawal and Nahmias 1997)
No competition exists as only one supplier is involved. It gives quantity discount from order consolidation and reduces order lead time and logistical lead time (Hahn et al. 1986; Bozarth et al. 1998)	Reduction in price is achieved through competition between suppliers. It gives better on-time delivery and higher volume flexibility (Ramasesh et al. 1991)
It is applicable where goodwill trust exists between buyer and supplier	Dual sourcing is always effective for low ordering cost and highly variable lead times (Ramasesh et al. 1991)
Low threat to loss of information	Since business data are shared among various suppliers, proper security measures should be taken

2.2 State-of-the-Art Literature Review of Supplier Selection Methods

The abundant work on supplier selection can be broadly classified into eight different categories as follows:

1. selection of supplier for single item or multiple items for deterministic or stochastic demand and supply;
2. selection of supplier for manufacturing industry;
3. selection of supplier for service industry;
4. selection of supplier with price–order quantity discount;
5. comparative analysis of single-sourcing and multiple-sourcing strategies;
6. decision support system (DSS) for supplier selection;
7. supplier selection for green supply chain; and
8. supplier selection for new product development.

In this chapter, literature review is conducted to find the followings:

1. to identify relevant criteria for supplier selection;
2. to identify different methods for supplier selection; and
3. to identify the trend of supplier selection methods.

Research work related to supplier selection is considerably very high. For instance, from www.sciencedirect.com, alone 13, 201 articles were found with the search word ‘supplier selection’ for publication 2009 onwards. About 100 research manuscripts are selected from peer-review journals from 1998 to 2012. Papers are selected based on the reputation of journal and citation of papers to find the most

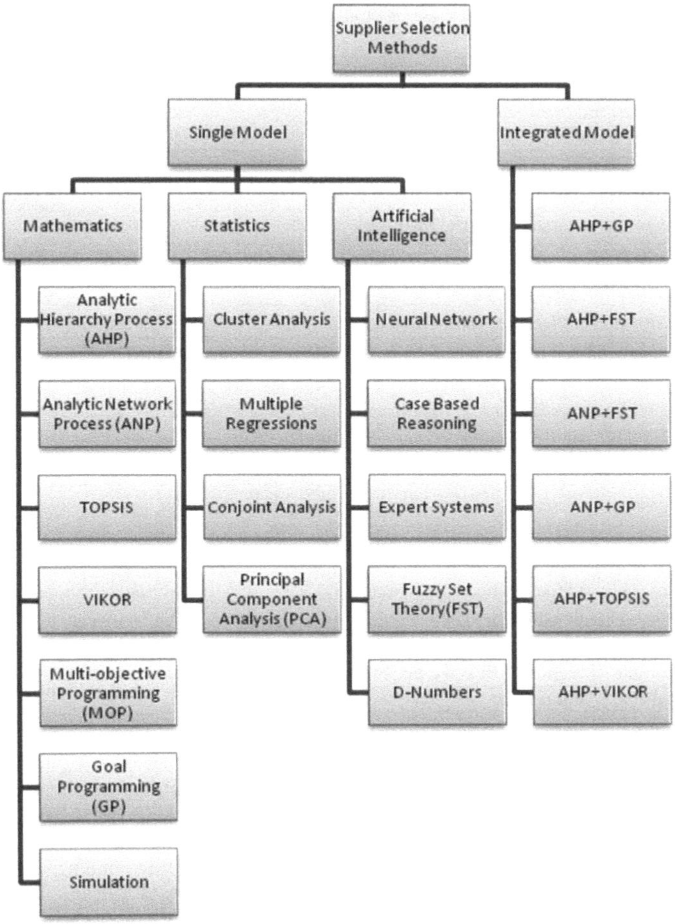

Fig. 2.1 Supplier selection methods

cited method(s) for supplier selection, recent trend of supplier selection, etc. Supplier selection methods are broadly classified into two categories—methods for single model and integrated model, as shown in Fig. 2.1. Single model is further classified into three subcategories—methods based on mathematics, statistics, and artificial intelligence. Integrated models usually combined with **linear program (LP)** or genetic algorithm (GA) or particle swarm optimization (PSO) to allocate order among multiple suppliers. Each method has certain limitations. Therefore, selection of an appropriate method always remains a daunting task for decision makers. AHP, ANP, and their integrated methods are mostly preferred by various researchers because of its simplicity and ability to solve complex problem. Figs. 2.2 and 2.3 clearly indicate such trend. Today, majority of the companies prefer to

Table 2.2 Supplier selection technique—a brief review

Sl. No.	Techniques	Authors
1.	Logarithmic least square method (LLSM) for fuzzy AHP	Van Laarhoven and Pedrycz (1983).
2.	Modified logarithmic least square method (MLLSM)	Wang et al. (2006a, b)
3.	Fuzzy least square priority method	Xu (1996).
4.	Lambda-Max method	Csutora and Buckley (2001)
5.	Eigenvector method	Wang et al. (2008a, b)
6.	Fuzzy preference programming	Mikhailov (2003)
7.	Extent analysis	Chang (1996)

Table 2.3 Various techniques for single-sourcing and multi-sourcing supplier selection

Single sourcing			Multi-sourcing	
Sl. No.	Methods	Remarks	Sl. No.	Methods
1.	Linear weighted point	Depends heavily on human judgments	1.	Mixed integer programming
2.	Categorical method	Depends heavily on human judgments	2.	Goal programming
3.	Cost ratio	Very complicated and needs more financial information	3.	Single- / multi-objective programming
4.	AHP	More accurate than any other method (Ghodsypour and O'Brien 1998)	4.	Multi-attribute utility theory and AHP; AHP-LP; AHP-GA; AHP and multi-objective possibilistic linear programming (AHP-MOPLP) etc

reduce supply base, and because of that, research trend on supplier selection is gradually moving from multiple supplier selection to single supplier selection. However, very less number of research papers has been identified on **supply base reduction** (SBR). Different criteria used for supplier selection methods are shown in Table 2.4. Cost, quality, and service are mostly used for traditional supplier selection process, for example, delivery time, on-time delivery, and delivery reliability. About 78 papers are analyzed thoroughly from 2005 to 2012 to find out the application of supplier selection methods in different industries, as shown in Fig. 2.4 (Table 2.5).

Study reveals that major contribution in research related to supplier selection is obtained from Taiwan, USA, Turkey, Iran, and China. Their cumulative research work related to supplier selection is about 69% of total research work. Both India and UK occupies the same position. This major contribution also inspired researchers to contribute more on supplier selection methods for electronics,

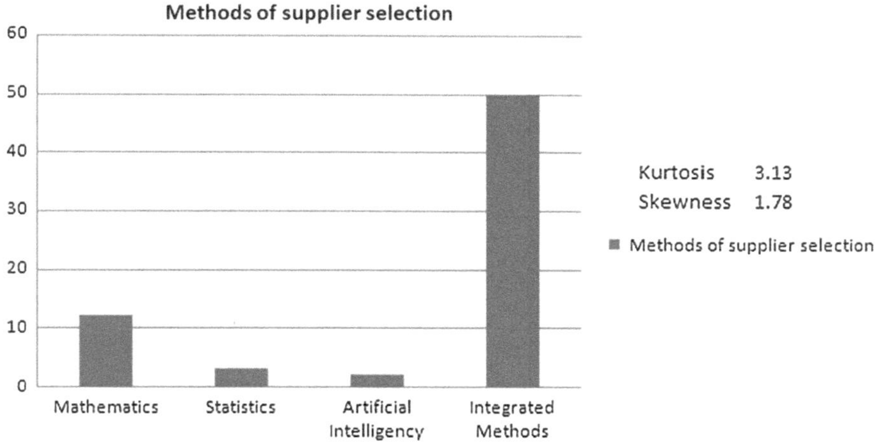

Fig. 2.2 Trend of supplier selection methods

Fig. 2.3 Distribution of research paper for supplier selection

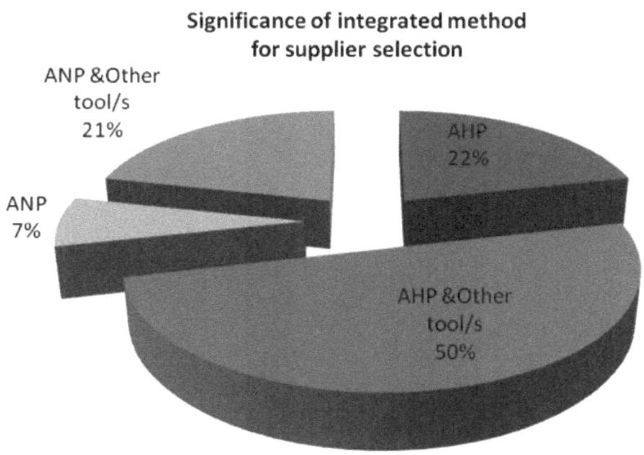

mechanical or manufacturing, and automobile industries, as shown in Fig. 2.4. Unfortunately, supplier selection for construction industries is yet to be explored.

2.3 Pareto Analysis of Supplier Selection Criteria

Dickson (1966) in his seminal work proposed twenty-three criteria for supplier selection. Based on the work of Dickson (1966) and Weber (1991), Pareto analysis is performed to find most cited criterion for supplier selection. Six criteria such as net price, delivery, quality, production facilities and capacity, geographic location, and technical capability are identified as the most cited criteria.

Table 2.4 A partial list of supplier selection criteria

Year	Author/s	Cost	Quality	Service	Delivery	Reputation	Environment	Logistical performance	Commercial structure	Production	Technology	Responsiveness	Supplier's profile	Risk factor	Reliability	Lead time	Flexibility	Supplier's willingness	R&D
1998	Ghodsypour, O'Brien	X	X	X															
2001	Eon-Kyung Lee, Sungdo Ha, and Sheung-Kown Kim,	X	X	X	X														
2001	Maggie C. Y. Tam, Rao Tummala	X	X																
2002	Robert Handfield, Steven V. Walton, Robert Sroufe, and Steven A. Melnyk						X												
2005	Ozden Bayazit and Birsen Karpak							X	X	X									
2006	Huan-Jyh Shyur and Hsu-Shih Shih	X	X		X						X	X							
2007	FU Yao and LIU Hongli	X	X		X														
2007	Felix T.S. Chan and Niraj Kumar	X	X	X									X	X					
2007	Weijun Xia and Zhiming Wu	X	X	X															
2007	Min Wu	X	X	X	X														
2007	Sanjay Jharkharia and Ravi Shankar	X	X			X													
2007	Cevriye Gencer and Didem Gürpinar		X							X									
2007	Ezgi Aktar Demirtas and Ozden Ustun	X	X	X															

(continued)

Table 2.4 (continued)

Year	Author/s	Cost	Quality	Service	Delivery	Reputation	Environment	Logistical performance	Commercial structure	Production	Technology	Responsiveness	Supplier's profile	Risk factor	Reliability	Lead time	Flexibility	Supplier's willingness	R&D
2008	Ali Kokangul and Zeynep Susuz	X			X														
2008	Jia-Wen Wang, Ching-Hsue Cheng, and Huang Kun-Cheng	X	X	X															
2008	Reuven R. Levary													X	X				
2008	Jing-Rung Yu and Chao-Chia Tsai	X	X	X	X		X												
2008	Ozan Cakir and Mustafa S. Canbolat	X																	
2008	Sung Ho Ha and Ramayya Krishnan		X													X			
2008	Amy H.I. Lee	X	X		X						X						X		
2008	Eleonora Bottani and Antonio Rizzi																	X	
2008	Rong-Ho Lin	X	X		X						X								
2008	Semih Önüt, Selin Soner Kara, and Elif Is_ik	X	X		X														
2008	Ozeden Ustun and Ezgi Aktar Demirtas	X	X		X														
2009	Wann-Yih Wu et al.	X	X																
2009	Chia-Wei Hsu, Allen H. Hu		X																X

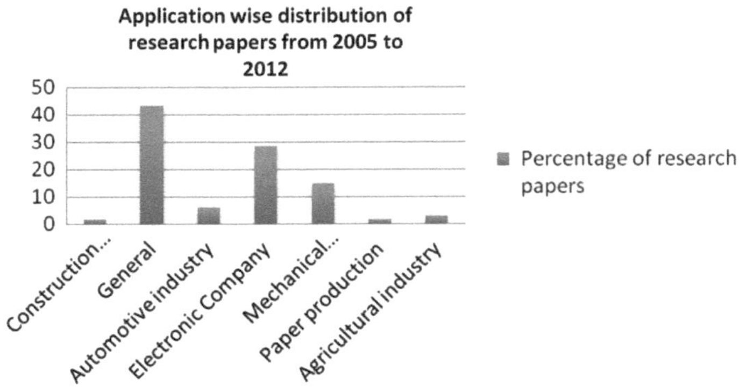

Fig. 2.4 Application of supplier selection methods to different industries

2.4 Stages of Procurement

Supplier selection is the process by which suppliers are reviewed, evaluated, and chosen to become part of the company's supply chain (Sanayei et al. 2010). The overall objective of the supplier selection process is as follows (Chena eta al 2006):

1. to reduce the procurement risk;
2. to maximize the overall value of purchase; and
3. to build the closeness and long-term relationships between buyers and suppliers.

Supplier selection is not a mere clerical issue or a mere optimization problem. Supplier selection is a strategic issue of any business because of the following reasons:

1. Procurement is considered as value addition process to supply chain.
2. Active supplier involvement can enhance efficiency and effectiveness of supply chain.
3. Short product life cycle and rapid product innovation give more emphasizes on integration of material and information flows, both internally and externally.

Supplier selection process consists of four stages—problem definition; formulation of attributes; qualification of potential suppliers; and the final selection of best suppliers (De Boer et al. 2001). A generalized **procurement cycle** can be considered that consists of the following stages:

1. Recognition of need: Identify the demand of product.
2. Specification: Identify part/assembly/raw material specifications.
3. Make or buy decision: It is one of the most crucial stages of procurement cycle to think over about source materials, goods, price, etc. Usually, a company is supposed to take, make, or buy decision for the following reasons:

Table 2.5 Application area wise distribution of research paper from 2005 to 2012

Year	Author(s)	Application areas
2005	Hong et al.	Agricultural industry in Korea
2005	Chen et al.	Electronic components
2005	Bayazit and Karpak	Construction company
2006	Kubat and Yuce	General
2006	Mouli et al.	General
2006	Sarfaraz and Balu	General
2006	Chen et al.	High tech manufacturing
2006	Shyura and Shih	Local Taiwanese company
2007	Gencer and Gu¨rpinar	Electronic company
2007	Demirtas and Ustun	Refrigerator producers
2007	Reza Farzipoor Saen	General
2007	Che et al.	Semiconductor industry
2007	Mehdizadeh and Moghaddam	General
2007	Guo et al.	General
2007	Yao and Hongli	Information & Mgmt Sys outsourcing
2007	Fayez et al.	General
2007	Huang and Keskar	PC manufacturer
2007	Amid et al.	General
2007	Li et al.	General
2007	Chan and Kumar	Manufacturing company
2007	Xia and Wu	General
2007	Min Wu	General
2007	Guan et al.	General
2008	Kokangul and Susuz	Automotive industry
2008	Rong-Ho Lin	General
2008	Moghadam et al.	General
2008a, b	Wang et al.	Lithium-ion battery
2008	Che and Wang	PDA
2008	Hong and Ha	Agricultural industry
2008	Reuven R. Levary	Manufacturing company
2008	Yu and Tsai	Semiconductor industry
2008	Lin and Chang	Manufacturing company
2008	Wan Lung Ng	General
2008	Amin and Razmi	ISP
2008	Wu et al.	TFT-LCD industry
2008	Chou and Chang	IT Industry
2008	Ha and Krishnan	Automobile industry
2008	Lee et al.	TFT-LCD industry
2008	Amy H.I. Lee	TFT-LCD industry
2008	Önüt et al.	Telecommunication industry

(continued)

Table 2.5 (continued)

Year	Author(s)	Application areas
2008	Zhang et al.	General
2008	Bottani and Rizzi	Manufacturing company
2008	Rhee et al.	Manufacturing company
2008	Wu and Olson	General
2008	Yang et al.	Electronic manufacturing company
2008	Ustun and Dem˙ırtas	Refrigerator manufacturing
2009	Amy H.I. Lee	TFT-LCD industry
2009	Wu et al.	Notebook manufacturer
2009	Hsu and Hu	Electronic manufacturing company
2009	Wang et al.	Notebook manufacturer
2010	Wen-Pai Wang	Electronic manufacturing company
2010	Wu et al.	General
2010	Sanayei et al.	Automobile industry
2010	Tadeusz Sawik	General
2011	Bilsel and Ravindran	General
2011	Tadeusz Sawik	General
2011	Amid et al.	General
2011	Selin Soner Kara	Paper production
2012	Erdem and Göçen	White goods manufacturer
2012	Shaw et al.	Garment manufacturing
2012	Mukherjee and Kar	General
2012	Jin Wang	General
2012	Riedl et al.	General
2012	Bruno et al.	General
2012	Choudhary and Shankar	General
2012	Parthiban et al.	Automotive industry

   1. sudden increase in procurement cost;
   2. need for design secrecy;
   3. lack of specific technical competency of suppliers in supply base;
   4. poor services of existing suppliers; and
   5. Unpredictable deterioration of existing supplier's performance, etc.

4. Source Identification: Prepare a supply base as per requirement.
5. Source selection: Organization has to think about single sourcing or optimal number of sourcing as per the goal of organization. Usually, it consists of four stages:

   1. Select criteria to consider palpable and non-palpable issues of supplier selection.
   2. Select appropriate method for supplier selection.

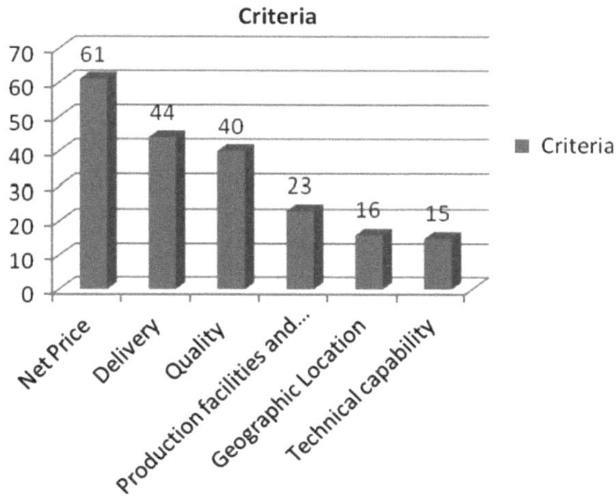

Fig. 2.5 Most cited criteria for supplier selection

3. Select supplier(s) as per the unanimous preference of decision makers.
4. Allocate order to selected suppliers as per mathematical model.

For single sourcing, last stage is not considered as entire order is allocated to single/best supplier.

6. Contracting: Placed order to selected suppliers.

In brief, supplier selection procedure can be classified as preselection, selection, and post-selection.

Preselection and post-selection are highly subjective and varied from company to company as per their goal of procurement process.

2.5 Qualities of Good Supplier

Quality of good supplier is highly subjective as it varied from company to company, product to product, and process to process. It also depends on the type of sourcing decisions. Sourcing decisions are classified as follows:

1. consumable supplies;
2. production materials and components;
3. capital purchases (e.g., machinery);
4. intellectual property (e.g., software);
5. subcontractors; and
6. services.

Based on the above classifications, a generalized list of qualities of good suppliers can be mentioned as follows:

1. on-time delivery;
2. technical capabilities;

3. consistent quality;
4. reasonable low price;
5. good past performance record;
6. ability to maintain volume flexibility to withstand sudden variations in demand;
7. presales and post-sales support;
8. ability to provide his buyers tracking facilities to track the progress of supply. Such tracking process could enhance the reliability of supply;
9. industrial certifications such as ISO and TUV; and
10. proactive to develop a healthy relation with his buyers.

2.6 How to Prepare Supply Base?

A stable supply base could enhance availability of raw material/parts/assembly, increase buyers' bargain power, and increase the possibility to get best supplier(s) to develop a long-term relationship. Preparation of **supply base** is not a much discussed topic in supplier selection literature as major focus is given to selection and evaluation of supplier(s). Often, it is considered as a preprocess of supplier selection. Author strongly suggests that due care should be given to prepare supply base as poor supply base gives low probability of selecting good supplier(s) even if very supplier selection method is used. Following are some of the criteria for preparing supply base:

1. reputation and industrial certification of suppliers;
2. availability of past performances of supplier with authenticated documents;
3. availability of well-documented product catalog. For raw material, a well-documented test report which contains chemical and physical properties should be provided.
4. Ability to provide goods as per delivery due date. Usually, it varies from company to company.
5. For new product, technical know-how of supplier should be verified. At the same time, availability of technical equipments of the supplier should be verified.

2.7 Supplier Selection for Mass Customized System

In Chap. 3, a detailed discussion is given for **customized production system**, **postponement,** and **CODP**. **Mass customization** is a strategy to manufacture customized product from standard product with near mass production efficiency. Mass customization can be broadly classified as ***assembly to order (ATO), build to order (BTO), engineer to order (ETO),*** *and* ***make to stock (MTS)*** which is commonly used for standard product to reduce customer's waiting time. Literature review reveals ample work on MTS but very few works on ATO/BTO/ETO. In this book, main focus is given to supplier selection methods, development of hybrid model for supplier selection, development of mathematical model for supplier selection for ATO system, and design of decision support system for sustainable supplier selection and strategic sourcing.

2.8 Hybrid Methods for Supplier Selection

In this chapter, three hybrid methods are proposed for supplier evaluation, selection, and order allocation problem. In the first method, modified extent fuzzy AHP is used to consider tangible and intangible criteria for supplier selection, and for **order allocation,** GA is used. In the second method, fuzzy AHP-TOPSIS is used to consider palpable and non-palpable criteria for supplier selection, and order allocation GA is used.

2.8.1 Modified Extent Fuzzy AHP and GA (MEFAHP-GA)

Chang's (1996) extent analysis is based on the following steps:

1. If $M_{g_i}^{i}$ are the triangular fuzzy numbers (TFNs) where g_i is the goal set (i = 1, 2, 3…m),
   the fuzzy extent value S_i with respect to the ith criterion is defined as

$$S_i = \sum_{i=1}^{m} M_{g_i}^{i} \otimes [\sum_{i=1}^{n}\sum_{i=1}^{m} M_{g_i}^{i}]^{-1} \tag{2.1}$$

   where $M_{g_i}^{i} = \left(\sum_{i=1}^{m} l \cdot \sum_{i=1}^{m} m \cdot \sum_{i=1}^{m} u\right)$
   where l is the lower limit value, m is the most promising value, and u is the upper limit value.
   and

$$[\sum_{i=1}^{n}\sum_{i=1}^{m} M_{g_i}^{i}]^{-1} = \left\{\frac{1}{\sum_{i=1}^{n} u_i}, \frac{1}{\sum_{i=1}^{n} m_i}, \frac{1}{\sum_{i=1}^{n} l_i}\right\} \tag{2.2}$$

2. The degree of possibility of $M_2 \geq M_1$ is given by $V(M_2 \geq M_1)$
   where

$$V(M_2 \geq M_1) = \begin{cases} 1, & if\ m_2 \geq m_1 \\ 0, & if\ l_1 \geq u_2 \\ \frac{l_1 - u_2}{(m_2 - u_2) - (m_1 - l_1)}, & otherwise \end{cases}$$

If d is the highest intersection point of μ_{M_1} and μ_{M_2} as shown in Fig. 2.6,

$$d(A_i) = \min V(S_i \geq S_k) \quad for \quad k = 1, 2, 3, 4, 5 \ldots n;\ k \neq i$$

The weight vector is $W^* = (d(A_1), d(A_2), d(A_3), \ldots, d(A_n))^T$
The normalized weight vector is $W = \frac{W^*}{\sum_{l=1}^{n} d(A_i)}$.

TFNs are used for fuzzy comparisons, as shown in Table 2.6.

Reason behind the popularity of extent fuzzy AHP is its computational simplicity. However, it is unable to find the true weights from fuzzy comparison matrix. Wang et al. (Wang et al. 2008a, b) pointed out that Eq. (2.2) should be modified to find the true fuzzy extent value of ith criteria. This method is known as modified extent fuzzy AHP (MEFAHP).

$$\left[\sum_{i=1}^{n}\sum_{i=1}^{m} M_{g_i}^{i}\right]^{-1}$$

$$= \left(\frac{\sum_{j=1}^{n} l_{ij}}{\sum_{j=1}^{n} l_{ij} + \sum_{k=1,k\neq i}^{n}\sum_{j=1}^{n} u_{kj}}, \frac{\sum_{j=1}^{n} m_{ij}}{\sum_{k=1}^{n}\sum_{j=1}^{n} m_{kj}}, \frac{\sum_{j=1}^{n} u_{ij}}{\sum_{j=1}^{n} u_{ij} + \sum_{k=1,k\neq i}^{n}\sum_{j=1}^{n} l_{kj}}\right)$$

where i = 1,2,3…n.

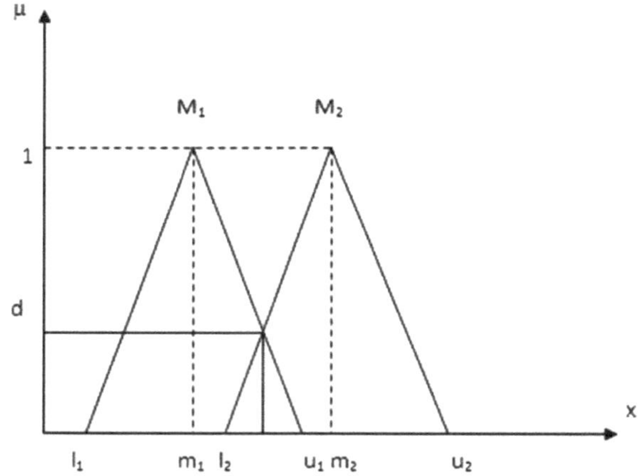

Fig. 2.6 Intersection of two TFNs (from Zhu et al. 1999; with kind permission from Elsevier Limited)

Table 2.6 Fuzzy TFN values (from Tolga et al. 2005; with kind permission from Elsevier Limited)

Linguistic values	Fuzzy numbers
Equal	(1,1,1)
Weak	(2/3,1,3/2)
Fairly strong	(3/2,2,5/2)
Very strong	(5/2,3,7/2)
Absolute	(7/2,4,9/2)

Wang et al. (2008a, b) further proposed that extent analysis method may be considered as the method for showing how bigger one a decision is than the others in fuzzy comparison and there is chances of loosing of information when it assigns irrational zero to some important criteria or subcriteria. Linear program (LP), integer program (IP), goal program, etc., can be used alone to consider limitations of supplier(s) in supplier selection problem. However, these methods cannot consider qualitative criteria for supplier selection. Therefore, combination of AHP-GA or AHP-LP is a better choice. Extent fuzzy AHP alone is more suitable for single sourcing where best supplier is capable enough to fulfill the entire demand. The following steps are used to combine extent fuzzy AHP with GA.

Step 1. Define goal or objective of the problem.
Step 2. Select criteria for selecting suppliers.
Step 3. Select fuzzy membership function for fuzzy comparison matrix.
Step 4. Find priority of suppliers by extent fuzzy AHP.
Step 5. Form objective function and constraints.
Step 6. Use genetic algorithm to solve single-objective constrained objective function.

Order allocation to selected suppliers with GA is discussed in detail in the next section mentioned below.

2.8.2 *Fuzzy TOPSIS-MOGA*

The following steps are maintained to rank suppliers from a predefined supply base:

Step 1: TFNs, $\tilde{a}_{ij}$, are used to find suitability of each alternatives w.r.t criteria.
Step 2: Since TFNs are already normalized, no need for normalization. Calculate the weighted normalized fuzzy decision matrix, $\tilde{U}$.

$$\widetilde{U} = [\tilde{u}_{ij}]_{n\times j} \quad i = 1,2,3\ldots n \quad j = 1,2,\ldots j$$

$$\tilde{u}_{ij} = \tilde{a}_{ij} \times w_i$$

where w_i is the set of weight of each criterion derived by analytic hierarchy process (AHP).
Step 3: Identify FPIS (fuzzy positive ideal solution) and FNIS (fuzzy negative ideal solution). Calculate the distance of each alternative from FPIS and FNIS.
Step 4: Calculate the closeness coefficient of each alternative, $CC_{i.}$
where $CC_j = \frac{D_j^-}{D_j^+ + D_j^-}$
Step 5: Rank suppliers based on higher value of $CC_{i.}$.

2.8.3 Multi-Objective Model for Supplier Selection

A multi-objective order allocation model is developed with five objective functions and three constraints. The following assumptions are considered to develop order allocation model

Assumptions

1. Selected suppliers will supply only one item.
2. No quantity discount is considered.
3. No shortage of item is allowed for any supplier.
4. Deterministic constant demand is considered.

C_i	Procurement cost of per ton of coal from ith supplier
TC_i	Transportation cost of per ton of coal from ith supplier
CC_i	Closeness coefficient of ith supplier obtained from fuzzy TOPSIS
α_i	Reliability of ith supplier
X_i	Order quantity to ith supplier
LD_i	Percentage of late delivery from ith supplier
β_i	Percentage of coal contains 15–18% of ash in per ton received from ith supplier
γ_i	Percentage of coal contains 15–16% of moisture in per ton received from ith supplier
H	Handling cost per ton

Order allocation model:

Total cost of purchase (TCP) consists of purchase, transportation, order/setup, and holding cost. Order/setup cost is neglected in this mathematical model, and material handling cost is considered as holding cost.

Minimize total cost of purchase (TCP): $\sum_{i=1}^{n} C_iX_i + \sum_{i=1}^{n} TC_iX_i + H\sum_{i=1}^{n} X_i$

Second objective function in our mathematical model is similar to Ghodsypour and O'Brien (1998). However, their proposed objective function is modified as total value of reliable purchase (TVRP). Reliability of supply of each supplier is calculated from past performance data of supplier.

Maximize total value of reliable purchase (TVRP): $\sum_{i=1}^{n} \alpha_i CC_i X_i$

Third objective function is to mitigate supply risk. Fourth and fifth objective functions are to maintain desired quality level. In any cement company, every lot is accepted based on two quality parameters—ash content and surface moisture content.

Minimize number of late deliveries: $\sum_{i=1}^{n} LD_i X_i$

Minimize amount of rejected lot based on ash content: $\sum_{i=1}^{n} (1-\beta_i) X_i$

Minimize amount of rejected lot based on moisture content: $\sum_{i=1}^{n} (1-\gamma_i) X_i$

Constraints for supplier selection:

Supplier capacity constraint, minimum order quantity to fulfill demand constraint, and cost or budgetary constraint are some of the most significant constraints of order allocation model (Kumar et al. 2004; Ghodsypour and O'Brien 1998). The following constraints are considered to optimize above five objective functions:

Capacity constraint: $X_i \leq V_i$ for $i = 1, 2, 3\ldots n$

Demand constraint: $\sum_{i=1}^{n} X_i = D$

Cost constraint: $\sum_{i=1}^{n} C_i X_i \leq B$

Non-negativity constraint: $X_i \geq 0$ for $i = 1, 2, 3\ldots n$

The proposed model is solved by using MATLAB R 2009a and run it on a personal computer intel(R) Core(TM) 2 Duo 2.00 Ghz. This integrated model consists of four stages—preprocessing, supplier selection, order quantity

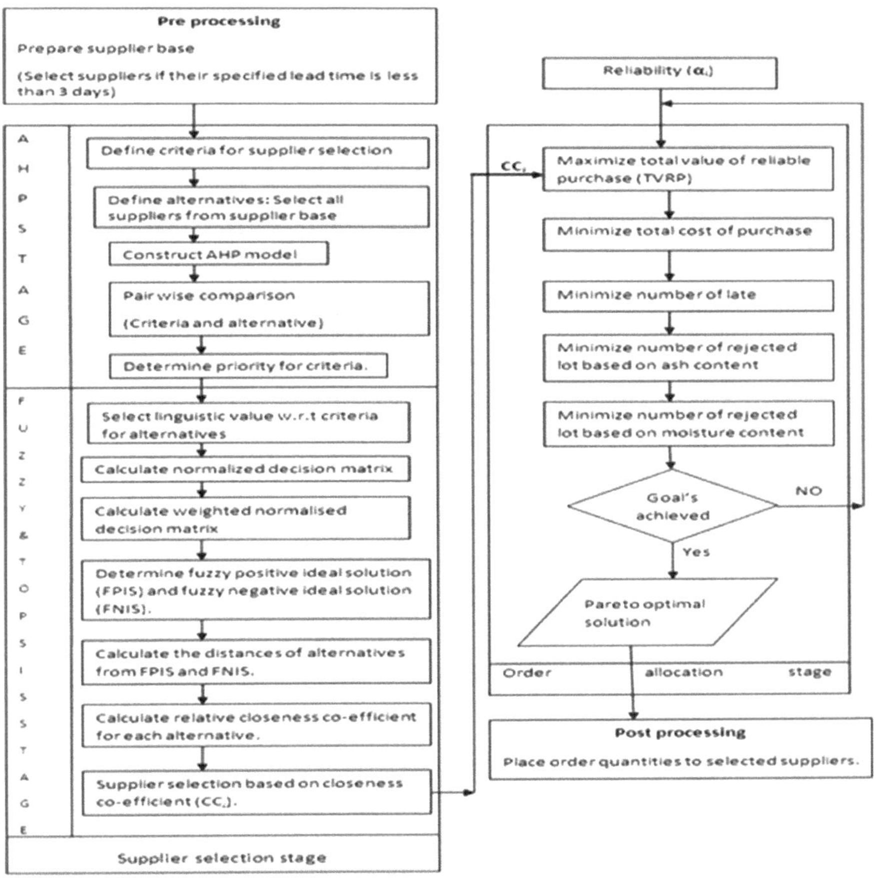

Fig. 2.7 Integrated model of f-TOPSIS-MOGA

calculation, and post-processing. In preprocessing, only preparation of supply base is considered. A supply pool is built initially with predefined supply lead time. In supplier selection stage, suppliers are selected from supply base by combined fuzzy TOPSIS and approval status proposed by Chen et al. (2006). Third stage uses genetic algorithm to optimize multi-objective, and finally, order quantities are selected from Pareto-optimal solutions. In the last stage, orders are placed to selected suppliers. Flowchart of the integrated approach is shown in Fig. 2.7.

2.8.4 Case Study

High initial investment, lack of resources, land acquisition problem for expansion, and long waiting time to get desired return on investment are some of the major barrier for cement industries. Moreover, full capacity utilization of cement plants is highly influenced by demand of realty sector. About 67% of the total production of cement is used in housing sector, 13% is used in commercial construction, 11% is used in infrastructure project, and only 9% is used in industrial construction. India

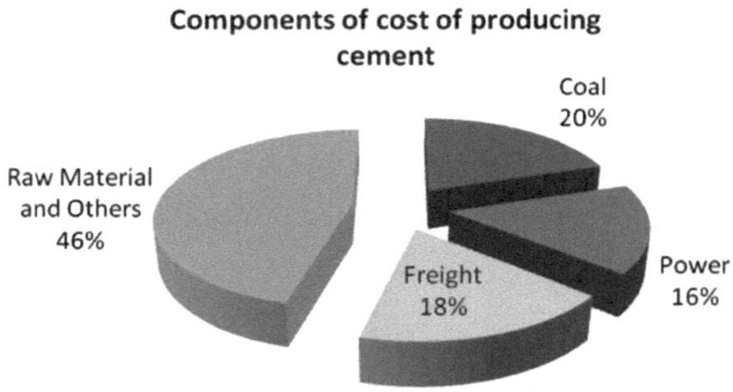

Fig. 2.8 Cost of producing cement

Table 2.7 Priority of supplier selection criteria

Criteria	Priority	Inconsistency
Quality(C_1)	0.3132	0.0935
Price(C_2)	0.0819	
Capacity(C_3)	0.0819	
Location(C_4)	0.5230	

Table 2.8 TFN values

Linguistic values	Fuzzy numbers
Very low (VL)	(0,0,0.2)
Low(L)	(0,0.2,0.4)
Medium (M)	(0.2,0.4,0.6)
High(H)	(0.4,0.6,0.8)
Very High(VH)	(0.6,0.8,1)
Excellent	(0.8,1,1)

is the second largest producer of cement after China. Indian cement industry is basically oligopolistic in nature with more than 160 companies scattered all over India. Northern, eastern, southern, western, and central are the five main regions responsible for cement production in India. Andhra Pradesh, Tamil Nadu, and Rajasthan are the main contributors to Indian cement industry. Till early 2000, Ordinary Portland Cement (OPC) was the main variety of cement in India. Since 2005, production of Portland Pozzolana Cement (PPC) was increased at the cost of production of OPC. Today, about 61% of total production is PPC. Total 20% of total cost is spent for procuring coal to produce cement, as shown in Fig. 2.8.

An ISO 9001:2000 certified company which is situated in north east is producing various grades of cement such as Ordinary Portland Cement (OPC) and Portland Pozzolana Cement (PPC). Limestone and coal are two important raw materials for cement. Gypsum is essential for OPC, and fly ash is essential for PPC. Company will select supply of coal if its ash content is 15–18% and surface moisture content is 15–16%. Moreover, company can wait maximum three days to get supply. Material handling cost comes to Rs. 350 per ton, and order/setup cost and other holding cost are negligible. Four criteria, namely quality, price, capacity, and location of the supplier, have been chosen to select suppliers as per the consensus of the decision maker's committee which encompasses senior members from finance, marketing, purchase, and sales department of the focal company.

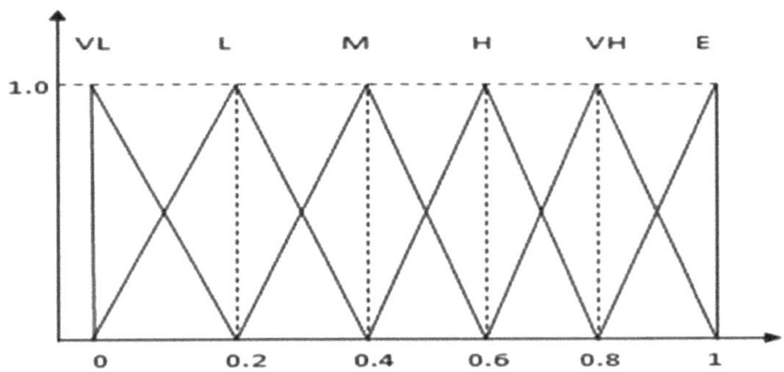

Fig. 2.9 Membership function of linguistic values

Table 2.9 Fuzzy evaluation matrix of alternatives

Alternative	C_1	C_2	C_3	C_4
A_1	High	Low	Medium	Medium
A_2	Medium	Medium	High	Very high
A_3	Very high	Low	High	Excellent
A_1	(0.4,0.6,0.8)	(0,0.2,0.4)	(0.2,0.4,0.6)	(0.2,0.4,0.6)
A_2	(0.2,0.4,0.6)	(0.2,0.4,0.6)	(0.4,0.6,0.8)	(0,0,0.2)
A_3	(0.6,0.8,1)	(0,0.2,0.4)	(0.4,0.6,0.8)	(0.8,1,1)
Weight	0.3132	0.0819	0.0819	0.5230

Table 2.10 Weighted evaluation for three suppliers

Alternative	C_1	C_2	C_3	C_4
A_1	(0.125,0.188,0.251)	(0,0.016,0.033)	(0.016,0.033,0.049)	(0.105,0.209,0.314)
A_2	(0.063,0.125,0.188)	(0.016,0.033,0.049)	(0.033,0.049,0.066)	(0,0,0.105)
A_3	(0.188,0.251,0.3132)	(0,0.016,0.033)	(0.033,0.049,0.066)	(0.418,0.523,0.523)
A^+	(1,1,1)	(0,0,0)	(1,1,1)	(0,0,0)
A^-	(0,0,0)	(1,1,1)	(0,0,0)	(1,1,1)

Pairwise comparison value for each criterion is obtained from each decision maker. After that, analytic hierarchy process (AHP) is used to calculate priority of each criterion, as shown in Table 2.6, and linguistic values are shown in Table 2.7 and in Fig. 2.9.

As stated above, detail calculation is shown in Tables 2.8, 2.9, 2.10, and 2.11. Supplier performance data are shown in Table 2.12.

All three suppliers can be accepted with low risk to supply coal.

Multi-objective functions for supplier selection:

In this problem, a linear total cost function (TC(Q)=a + bQ) is considered for all three suppliers for simplicity. As shown below, a **nonlinear integer function** is developed for total cost of purchase which is to be minimized, as shown in Fig. 2.10.

Minimize **total cost of purchase (TCP)**: $(3099 + 10 \times 1) \times 1 + (3100 + 10 \times 2) \times 2 + (3102 + 10 \times 3) \times 3$

Maximize **total value of reliable purchase (TVRP)**: $0.4674 \times 1 + 0.4982 \times 2 + 0.4133 \times 3$

Minimize delay in supply: $0.1 \times 1 + 0.15 \times 2 + 0.2 \times 3$

Quality:

1. Minimize defects to maintain permissible ash content in supply: $0.2 \times 1 + 0.25 \times 2 + 0.3 \times 3$
2. Minimize defects to maintain permissible moisture content in supply: $0.15 \times 1 + 0.2 \times 2 + 0.2 \times 3$

Subject to

Demand constraint: $\times 1 + \times 2 + \times 3 = 8000$

Production constraint: $\times 1 \leq 4000; \times 2 \leq 3000; \times 3 \leq 3000$

Budget constraint: $(2000 + 10 \times 1) \times 1 + (2000 + 10 \times 2) \times 2 + (2000 + 10 \times 3) \times 3 \leq 30000000$

Table 2.11 Fuzzy TOPSIS result

Alternatives	D_i^+	D_i^-	CC_i
A_1	2.0282	2.0092	0.4976
A_2	1.9228	2.1201	0.5244
A_3	2.2134	1.8052	0.4492

Table 2.12 Approval status

Closeness coefficient (CC_i)	Assessment status
$CC_i \in [0,0.3)$	Rejected
$CC_i \in [0.3,0.5)$	Recommended with high risk
$CC_i \in [0.5,0.7)$	Recommended with low risk
$CC_i \in [0.7,0.9)$	Approved
$CC_i \in [0.9,1.0]$	Highly recommended

Table 2.13 Supplier performance data

Sl No.	% failure rate of supply (f)	Reliability ($\alpha = 1 - f$)	Capacity (ton)	Total cost (Rs/ton)	Transportation cost (Rs/ton)	Quality	% late delivery	
						% of coal contains 15–18% ash in per ton	% of coal contains 15–16% moisture in per ton	
1.	6	0.94	4000	2000 + 10*Q	749	0.8	0.85	0.1
2.	5	0.95	3000	2000 + 10*Q	750	0.75	0.8	0.15
3.	8	0.92	3000	2000 + 10*Q	752	0.7	0.8	0.2

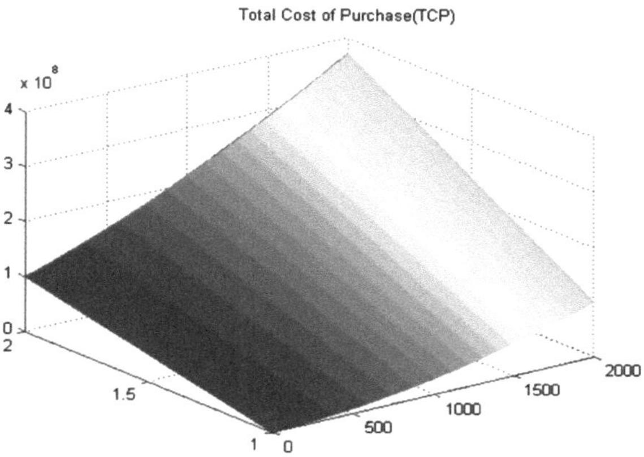

Fig. 2.10 Objective function for total cost of purchase (TCP)

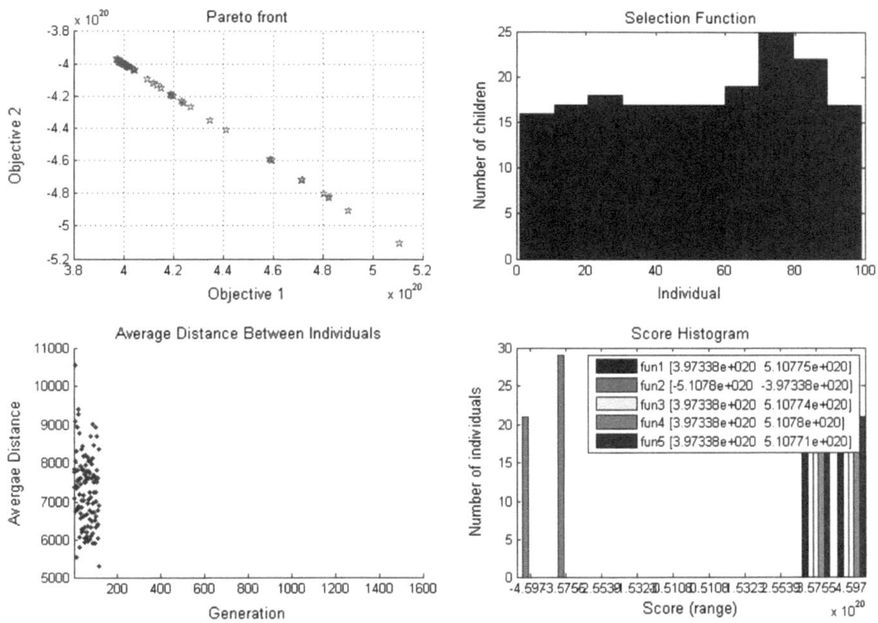

Fig. 2.11 Result of multi-objective GA

The multi-objective problem consists of nonlinear objective function with one nonlinear constraint. It cannot be solved with MATLAB GA solver. **Penalty function** approach is considered to convert constraint optimization problem to **unconstrained optimization** problem. Multi-objective GA finds multiple and diverse Pareto-optimal (or near Pareto-optimal) solutions in a single simulation run. Therefore, it is important to choose a particular solution from a set of nondominated solutions. A discontinuous Pareto front is obtained, shown in Fig. 2.11.

The number of points on the Pareto front was 50. The **average distance measure** of the solutions on the Pareto front was 0.0574885. The **spread measure** of the Pareto front was 0.0700832. Finally, ordered quantities to three suppliers are {2666, 2666, 2668}.

2.9 Conclusion

Supplier selection is not a mere clerical process. It encompasses several palpable and nonpalpable criteria. It is a multi-criteria-based optimization process. Effective selection of supplier could reduce uncertainty of availability of raw material, assure quality throughout the supply chain, reduce upstream supply chain risk, and, finally, reduce cost of manufacturing of product. About 70% of total cost is usually spent in procurement. Thus, procurement is most important for any company. In this chapter, several methods and criteria are mentioned through rigorous literature survey. Most cited criteria are identified with latest trend of supplier selection. As per the latest trend, two methods are discussed in detail with a case study. This chapter discusses in detail supplier pool preparation, supplier selection, evaluation, and order allocation with the above two methods.

Majority of the supplier selection models proposed by different researchers are for electronics industries, automobile industries, etc. In this chapter, an attempt has been made to prepare mathematical model to select suppliers for cement industries. Cement is a localized product and needs some extra constraints that are required to select and allocate order to suppliers. Proposed model in this chapter is prepared accordingly. It is pertinent to mention that application of the proposed models is not limited to the cement industries alone. It can be used for any industry with simple modification.

Total value of purchase is commonly used to allocate order to selected suppliers. In this chapter, total value of reliable purchase (TVRP) is considered instead of total value of purchase (TVP) to reduce the upstream supply chain risk, if any. TVRP is a weighted nonlinear objective function which is prepared with the priority obtained from the proposed MCDA tools. Industries such as cement and R&D usually face high risk to manufacture their product within due time. In such cases, TVRP should be used instead of TVP.

References

Agrawal N, Nahmias S (1997) Rationalization of the supplier base in the presence of yield uncertainty. Prod Oper Manage 6(3):291–308

Amid A, Ghodsypour SH, O'Brien C (2007) A weighted additive fuzzy multi-objective model for the supplier selection problem under price breaks in a supply chain. Int J Prod Econ. doi:10.1016/j.ijpe.2007.02.040
Amid A, Ghodsypour SH, O'Brien C (2011) A weighted max–min model for fuzzy multi-objective supplier selection in a supply chain. Int J Prod Econ 131:139–145
Amin SH, Razmi J (2008) An integrated fuzzy model for supplier management: a case study of ISP selection and evaluation. Expert Syst Appl. doi:10.1016/j.eswa.2008.10.012
Bayazita O, Karpak B (2005) An AHP application in vendor selection. In: Proceedings of ISAHP Honolulu, Hawaii, 2005 July 8–10
Bilsel RU, Ravindran A (2011) A multiobjective chance constrained programming model for supplier selection under uncertainty. Transp Res Part B 45:1284–1300
Bottani E, Rizzi A (2008) An adapted multi-criteria approach to suppliers and products selection—an application oriented to lead-time reduction. Int J Prod Econ 111:763–781
Bozarth C, Handfield R, Das A (1998) Stages of global sourcing strategy evolution: An exploratory study. J Oper Manage 16(2–3):241–255
Bruno G, Esposito E, Genovese A, Passaro R (2012) AHP-based approaches for supplier evaluation: problems and perspectives. J Purchasing Supply Manage 18:159–172
Cakir O, Canbolat MS (2008) A web-based decision support system for multi-criteria inventory classification using fuzzy AHP methodology. Expert Syst Appl 35:1367–1378
Chan FTS, Kumar N (2007) Global supplier development considering risk factors using fuzzy extended AHP-based approach. Omega 35:417–431
Chang DY (1996) Application of The Extent Analysis Method of Fuzzy AHP. Eur J Oper Res 95:649–655
Che ZH, Wang HS (2008) Supplier selection and supply quantity allocation of common and non-common parts with multiple criteria under multiple products. Comput Ind Eng 55:110–133
Che ZH, Wang HS, Sha DY (2007) A multi-criterion interaction-oriented model with proportional rule for designing supply chain networks. Expert Syst Appl 33:1042–1053
Chen KL, Chen KS, Li RK (2005) Suppliers capability and price analysis chart. Int J Prod Econ 98:315–327
Chen CT, Lin CT, Huang SF (2006) A fuzzy approach for supplier evaluation and selection in supply chain management. Int J Prod Econ 102:289–301
Chou S-Y, Chang Y-H (2008) A decision support system for supplier selection based on a strategy-aligned fuzzy SMART approach. Expert Syst Appl 34:2241–2253
Choudhary D, Shankar R (2012) Joint decision of procurement lot-size, supplier selection, and carrier selection. J Purchasing Supply Manage. doi:10.1016/j.pursup.2012.08.002
Csutora R, Buckley JJ (2001) Fuzzy hierarchical analysis: the Lamda-Max method. Fuzzy Sets Syst 120:181–195
Dağdeviren M, Yavuz S, Kilinç N (2009) Weapon selection using the AHP and TOPSIS methods under fuzzy environment. Expert Syst Appl 36(4):8143–8151
De Boer L, Labro E, Molrlacchi (2001) A Review of methods supporting supplier Selection. Eur J Purchasing Supply Manage 7(2):75–89
Deb K (2001) Multi-objective optimization using evolutionary algorithms. John Willey and Sons, New York
Demirtas EA, Ustun O (2007) Analytic network process and multi-period goal programming integration in purchasing decisions. Comp Ind Eng. doi:10.1016/j.cie.2006.12.006
Dickson GW (1966) An analysis of vendor selection systems and decisions. J Purchasing 2(1):5–17
Erdem AS, Göçen E (2012) Development of a decision support system for supplier evaluation and order allocation. Expert Syst Appl 39:4927–4937
Faez F, Ghodsypour SH, O'Brien C (2007) Vendor selection and order allocation using an integrated fuzzy case-based reasoning and mathematical programming model. Int J Prod Econ. doi:10.1016/j.ijpe.2006.11.022
Gaither N (1996) Production and operations management. Duxbury Press, Florence, KY

Gencer C, Gürpinar D (2007) Analytic network process in supplier selection: a case study in an electronic firm. Appl Math Model 31:2475–2486

Ghodsypour SH, O'Brien C (1998) A decision support system for supplier selection using an integrated analytic hierarchy process and linear programming. Int J Prod Econ 56–57:199–212

Ghodsypour SH, O'Brien C (2001) The total cost of logistics in supplier selection, under conditions of multiple sourcing, multiple criteria and capacity constraint. Int J Prod Econ 73:15–27

Guan Z, Jin Z, Zou B (2007) A multi-objective mixed-integer stochastic programming model for the vendor selection problem under multi-product purchases. Inf Manag Sci 18(3):241–252

Guo M, Zhu J, Zhao X (2007) A Bi-level programming model for supplier selection in constructing logistics service supply chain. In: Proceedings of 2007 IEEE IEEM ISBN: 1-4244-1529-2/07

Ha SH, Krishnan R (2008) A hybrid approach to supplier selection for the maintenance of a competitive supply chain. Expert Syst Appl 34:1303–1311

Hahn CK, Kim KH, Kim JS (1986) Costs of competition: implications for purchasing strategy. J Purchase Mater Manag (Fall)

Handfield R, Walton SV, Sroufe R, Melnyk SA (2002) Applying environmental criteria to supplier assessment: A study in the application of the analytical hierarchy process. Eur J Oper Res 141:70–87

Hong G, Ha SH (2008) Evaluating supply partner's capability for seasonal products using machine learning techniques. Comput Ind Eng 54:721–736

Hong GH, Park SC, Jang DS, Rho HM (2005) An effective supplier selection method for constructing a competitive supply-relationship. Expert Syst Appl 28:629–639

Hsu C-W, Hu AH (2009) Applying hazardous substance management to supplier selection using analytic network process. J Clean Prod 17:255–264

Huang SH, Keskar H (2007) Comprehensive and configurable metrics for supplier selection. Int J Prod Econ 105:510–523

Jang J, Sun C, Mizutani E (2004) Neuro-Fuzzy and soft computing: a computational approach to learning and machine intelligence. Prentice-Hall, India

Jharkharia S, Shankar R (2007) Selection of logistics service provider: an analytic network process. Omega 35:274–289

Kelle P, Silver EA (1990) Decreasing expected shortages through order splitting. Eng Costs Prod Econ 19:351–357

Kokangul A, Susuz Z (2008) Integrated analytical hierarch process and mathematical programming to supplier selection problem with quantity discount. Appl Math Model. doi:10.1016/j.apm.2008.01.021

Kubat C, Yuce B (2006) Supplier selection with genetic algorithm and fuzzy AHP. In: Proceedings of the 5th international symposium on intelligent manufacturing systems, 29–31 May

Kumar M, Vrat P, Shankar R (2004) A fuzzy goal programming approach for vendor selection problem in a supply chain. Comput Ind Eng 46:69–85

Lee AHI (2008) A fuzzy supplier selection model with the consideration of benefits, opportunities, costs and risks. Expert Syst Appl. doi:10.1016/j.eswa.2008.01.045

Lee AHI (2009) A fuzzy supplier selection model with the consideration of benefits, opportunities, costs and risks. Expert Syst Appl 36:2879–2893

Lee E, Ha S, Kim S (2001) Supplier selection and management system considering relationships in supply chain management. IEEE Trans Eng Manage 48(3):307–317

Lee AHI, Kang H-Y, Chang C-T (2008) Fuzzy multiple goal programming applied to TFT-LCD supplier selection by downstream manufacturers. Expert Syst Appl. doi:10.1016/j.eswa.2008.08.044

Levary RR (2008) Using the analytic hierarchy process to rank foreign suppliers based on supply risks. Comput Ind Eng 55:535–542

Li GD, Yamaguchi D, Nagai M (2007) A grey-based decision-making approach to the supplier selection problem. Math Comp Model 46:573–581

Lin R-H (2008) An integrated FANP–MOLP for supplier evaluation and order allocation. Appl Math Model. doi:10.1016/j.apm.2008.08.021

Lin H-T, Chang W-L (2008) Order selection and pricing methods using flexible quantity and fuzzy approach for buyer evaluation. Eur J Oper Res 187:415–428

Mehdizadeh E, Moghaddam RT (2007) A Hybrid Fuzzy Clustering PSO Algorithm for a Clustering Supplier Problem. In: Proc. 2007 IEEE IEEM ISBN: 1-4244-1529-2/07

Mikhailov L (2003) Deriving priorities from fuzzy pair wise comparison judgements. Fuzzy Sets Syst 134:365–385

Moghadam MRS, Afsar A, Sohrabi B (2008) Inventory lot-sizing with supplier selection using hybrid intelligent algorithm. Appl Soft Comput 8:1523–1529

Mouli KVVC, Subbaiah KV, Rao KM, Acharyulu SG (2006) Particle swarm optimization approach for vendors selection. IE(I) J-PR 87:3–6

Moynihan GP, Saxena P, Fonseca DJ (2006) Development of decision support system for procurement operations. Int J Logistics Sys Manag 2(1):1–18

Mukherjee S, Kar S (2012) A three phase supplier selection method based on fuzzy preference degree. J King Saud Univ—Comp Inf Sci http://dx.doi.org/10.1016/j.jksuci.2012.11.001

Ng WL (2008) An efficient and simple model for multiple criteria supplier selection problem. Eur J Oper Res 186:1059–1067

Önüt S, Kara SS, Işik E (2008) Long term supplier selection using a combined fuzzy MCDM approach: a case study for a telecommunication company. Expert Syst Appl. doi:10.1016/j.eswa.2008.02.045

Opricovic S, Tzeng G (2004) Compromise solution by MCDM methods: a comparative analysis of VIKOR and TOPSIS. Eur J Oper Res 156:445–455

Parthiban P, Zubar HA, Garge CP (2012) A multi criteria making approach for suppliers selection. Procedia Eng 38:2312–2328

Ramasesh RV, Ord JK, Hayya JC, Pan AC (1991) Sole versus dual sourcing in stochastic lead time(s, Q) inventory models. Manage Sci 37(4):428–443

Riedla DF, Kaufmann L, Zimmermann C, Perols JL (2012) Reducing uncertainty in supplier selection decisions: antecedents and outcomes of procedural rationality. J Oper Manag. doi:10.1016/j.jom.2012.10.003

Saen RF (2007) Suppliers selection in the presence of both cardinal and ordinal data. Eur J Oper Res 183:741–747

Sanayei A, Mousavi SF, Yazdankhah A (2010) Group decision making process for supplier selection with VIKOR under fuzzy environment. Expert Syst Appl 37:24–30

Sarfaraz AR, Balu R (2006) An integrated approach for supplier selection. IEEE ISBN: 0-7803-9701-0/06

Sawik T (2010) Single vs. multiple objective supplier selection in a make to order environment. Omega 38:203–212

Sawik T (2011) Supplier selection in make-to-order environment with risks. Math Comp Model 53:1670–1679

Shaw K, Shankar R, Yadav SS, Thakur LS (2012) Supplier selection using fuzzy AHP and fuzzy multi-objective linear programming for developing low carbon supply chain. Expert Syst Appl 39:8182–8192

Shyur HJ, Shih HS (2006) A hybrid MCDM model for strategic vendor selection. Math Comput Model 44:749–761

Soner Kara S (2011) Supplier selection with an integrated methodology in unknown environment. Expert Syst Appl 38:2133–2139

Srinivas N, Deb K (1994) Mult-iobjective optimization using non-dominated sorting in genetic algorithms. Evol Comput 2(3):221–248

Tam MCY, Tummala VMR (2001) An application of the AHP in vendor selection of a telecommunications system. Omega 29(2):171–182

Tolga E, Demircan ML, Kahraman C (2005) Operating system selection using fuzzy replacement analysis and analytic hierarchy process. Int J Prod Econ 97:89–117

Ustun O, Demirtas EA (2008) An integrated multi-objective decision-making process for multi-period lot-sizing with supplier selection. Omega 36:509–521
Van der Rhee B, Verma R, Plaschka G (2008) Understanding trade-offs in the supplier selection process: the role of flexibility, delivery, and value-added services/support. Int J Prod Econ. doi:10.1016/j.ijpe.2008.07.024
Van Laarhoven PJM, Pedrycz W (1983) A fuzzy extension of Saaty's priority theory. Fuzzy Sets Syst 11:199–227
Wang W-P (2010) A fuzzy linguistic computing approach to supplier evaluation. Appl Math Model 34:3130–3141
Wang J (2012) Do firms' relationships with principal customers/suppliers affect shareholders' income? J Corp Finan 18:860–878
Wang YM, Parkan C (2006) Two new approaches for assessing the weights of fuzzy opinions in group decision analysis. Inf Sci 176:3538–3555
Wang YM, Elhag TMS, Hua ZS (2006) A modified fuzzy logarithmic least squares method for fuzzy analytic hierarchy process. Fuzzy Sets Syst 157(23):3055–3071
Wang YM, Luo Y, Hua ZS (2008a) On the extent analysis method for fuzzy AHP and its applications. Eur J Oper Res 186(2):735–747
Wang J-W, Cheng C-H, Cheng HK (2008b) Fuzzy hierarchical TOPSIS for supplier selection. Appl Soft Comput. doi:10.1016/j.asoc.2008.04.014
Wang S-Y, Chang S-L, Wang R-C (2009) Assessment of supplier performance based on product-development strategy by applying multi-granularity linguistic term sets. Omega 37:215–226
Weber CA, Current JR, Benton WC (1991) Vendor selection criteria and methods. Eur J Oper Res 50:2–18
Wu M (2007) Topsis-AHP simulation model and its application to supply chain management. World J Model Simul 3(3):196–201
Wu D, Olson DL (2008) Supply chain risk, simulation, and vendor selection. Int J Prod Econ 114:646–655
Wu WY, Shih H-A, Chan H-C(2008)The analytic network process for partner selection criteria in strategic alliances. Expert Sys Appl. doi:10.1016/j.eswa.2008.06.049
Wu W-Y, Sukoco BM, Li C-Y, Chen SH (2009) An integrated multi-objective decision-making process for supplier selection with bundling problem. Expert Syst Appl 36:2327–2337
Wu DD, Zhang Y, Wu D, Olson DL (2010) Fuzzy multi-objective programming for supplier selection and risk modeling: a possibility approach. Eur J Oper Res 200:774–787
Xia W, Wu Z (2007) Supplier selection with multiple criteria in volume discount environments. Omega 35:494–504
Xu R, Zhai X (1996) Fuzzy logarithmic least squares ranking method in analytic hierarchy process. Fuzzy Sets Syst 77:175–190
Yang JL, Chiu HN, Tzeng G-H, Yeh RH (2008) Vendor selection by integrated fuzzy MCDM techniques with independent and interdependent relationships. Inf Sci 178:4166–4183
Yao F, Hongli L (2007) Information systems outsourcing vendor selection based on analytic hierarchy process. IEEE ISBN: 1-4244-1312-5/07
Yu J-R, Tsai C-C (2008) A decision framework for supplier rating and purchase allocation: A case in the semiconductor industry. Comput Ind Eng 55:634–646
Zhang D, Zhang J, Lai K-K, Lu Y (2008) An novel approach to supplier selection based on vague sets group decision. Expert Syst Appl. doi:10.1016/j.eswa.2008.07.053

Chapter 3
Mass Customization

Mass customization combines economy of scale and economy of scope to enhance supply chain surplus with demand aggregation.

3.1 Introduction

Heterogeneous customer base, customer demand for configurable product, significantly shorter product life cycle, volatile market demand, and demand for customer responsiveness continuously pressurize companies how to satiate and retain potential customer. Mass customization strategies can be an obvious panacea for this issue. However, does mass customization pay? In the late 1960s and the early 1970s, buyers were tired of the poor quality, long customer lead time, high price, and monotonous variety of standard product. Increasing demand for product variety and limited manufacturing facilities compelled academician to think over alternative operations strategy and competitive priority (Skinner 1969). They realized that wining strategy of competitive market was moved from price-driven market to cost-, quality-, delivery-, and flexibility-driven market. In 1950, Alderson introduced the postponement. In 1965, first study of postponement was started at Buckling. During this phase, postponement got appreciation in the industry. However, it failed to win the corporate. In 1980s, Japanese firms first realized the importance of quality and flexibility within short period of time Japanese firms got monopoly in the world market. In 1987, Stan Davis introduced mass customization in his seminal work-Future Perfect. However, people gave scant attention as because they were preoccupied with quality issues. On the other hand, another concept popularly known as customer order decoupling point (CODP) was introduced by Bucklin [21] in 1965, and it got much attention during this period. In 1993, Pine published "Mass Customization: the New Frontier in Business Competition." With the advent of internet technology at the end of 1990s, much awaited process of customization started to gain its momentum.

We can classify companies as product-based and customer-based. In product-based company, standard products are manufactured irrespective of customer choice.

K. Mukherjee, *Supplier Selection*, Studies in Systems, Decision and Control 88,
DOI 10.1007/978-81-322-3700-6_3

In customer-based company, products are manufactured/assembled as per the need of individual customer. For example, Adidas develops running shoe equipped with magnetic sensors and a microprocessor to measure and control the compressive force for the wearer (Salvador et al. 2009). How to align product attribute as per the heterogeneous demand of customer? What would be the appropriate strategy to implement mass customization effectively? According to Salvador et al. (2009), a company must possess the following three fundamental capabilities to offer customized product to customer:

1. **Development of the solution space:** Identify the need of potential customers and render the information to develop/modify product characteristics. Create virtual prototype as per the "virtual concept testing." "Virtual concept testing" was implemented by Adidas to minimize redundant variants.
2. **Robust process design:** Reuse or recombine available resources to develop customized product. Extensive need of use FMS, process modularity, and adaptive human capital. Here, adaptive human capital means skilled/trained people who can take new tasks.
3. **Choice navigation:** Too many choices for product variants could create customer's confusion. Is properly known as "paradox of choice." Assortment matching is one of the effective approaches to map customer's need with available product variants.

3.2 Constraints of Mass Customization

The purpose of mass customization is to satisfy individual customer needs with near mass production efficiency (Tseng et al. 1996).Tseng and Jiao (2004) identified three aspects of customizability—design customizability, process customizability, and perceived value of customized product. Based on their work, FDPT constraints are indentified—financial constraints, design constraints, process constraints, and time constraints. In Table 3.1, common expectations of customer, constraints, and

Table 3.1 Common constraints and solutions of mass customization

Expectations of customers	Constraints	Solutions
Cost: lower is better	Financial constraints	Reduce the cost of procurement, production, and logistics as much as possible
Design flexibility and product variety: higher is better	Design constraints	Identify the need of customer and allow customer to work as co-designer. Such approach will integrate customer value in right amount with product
Process flexibility: higher is better	Process constraints	Standardize processes and use advanced manufacturing technology such as FMS to enhance the process flexibility
Time to offer customized product: lower is better	Time constraints	Reduce total lead time

their corresponding solutions are shown to implement effective strategies of mass customization.

3.3 Postponement

Significance of using postponement strategy is to reap benefits of leagile concept (Van Hoek 2000).The postponement strategy uses demand aggregation to reduce the difference between high demand and low demand (Cheng, p. 3). Effective implementation of postponement strategy can reduce process complexity, and demand and supply uncertainties. There are four basic categories of postponement strategies, namely pull postponement, logistics postponement, form postponement (Lee 1998), and price postponement (van Mieghem and Dada 1999). Production postponement, on the other hand, encompasses first three postponement strategies stated above (van Mieghem and Dada 1999). Bowersox and Closs (1996) stated three different postponement strategies—time postponement, place postponement, and form postponement.

3.4 Sourcing Postponement—A New Kind of Postponement Strategy

Procure when required

Sourcing postponement is useful for complex product design environment where complex customized product is prepared from base product by adding auxiliary components, as shown in Fig. 3.1. Company can purchase all components at a time to get price discount; however, it could create difficulties in inventory management. On the other hand, fluctuation of demand at the downstream could create stock out or over stock situation of components at middle stream or far upstream. In such situation, one strategic approach could be to purchase raw material/components from supplier and manufacture base product as per aggregate forecast and procure auxiliary components at the retailer site after receiving customer order. Such sharing of inventories at manufacturer and retailer site could reduce the risk of stock out/over stock, WIP inventory, and customer waiting time.

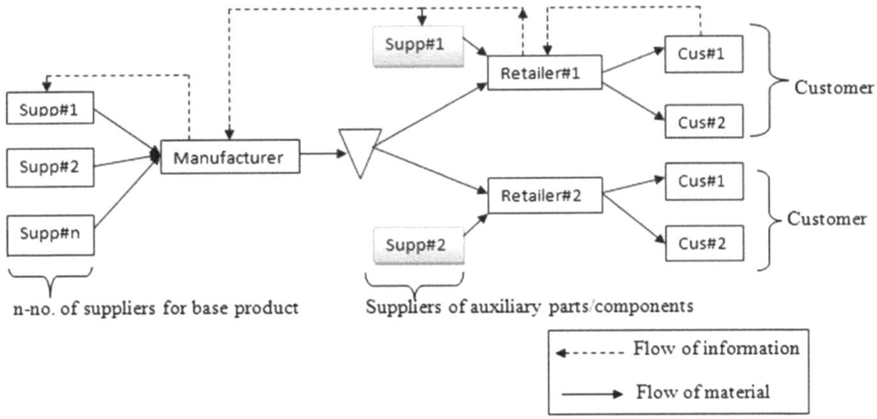

Fig. 3.1 Sourcing postponement

3.5 Advantages of Postponement Strategy

Both the philosophy of postponement and the principle of just in time (JIT) emphasize procurement strategies to mitigate upstream supply chain risk. Effective implement of postponement strategy could save time, improve quality, and reduce cost of product. Efficiency of postponement strategy depends solely on product family architecture (PFA) as this hierarchical structure of product family integrates economy of scales and economy of cost. Postponement strategy simplifies forecasting of product as product remains in generic form instead of finished form. Therefore, internal variability of product is drastically reduced without spoiling the external variability of product. Modular characteristics of product not only reduce the cost production but also enhance outsource capability of auxiliary processes and speed up the process of new product development. Van Hoek (2001) identified that postponement strategy reduces uncertainties by delaying finalization of products, increase product variety with product family architecture (PFA), and reduce complexity of supply chain. In traditional supply chain, lead time increases, variety brings more obsolescence risks and more uncertainties due to push-type supply chain (Van Hoek 2001). Integration of push-type and pull-type supply chain offers an alternative way to tackle such issues with product differentiation or delayed product finalization. However, product differentiation does not guarantee to mitigate all expected risks that may develop due product complexity. A large product variety may create several problems for the company. A well-defined product family architecture is highly desired in this regard. Modular approach for manufacturing can be expected as one of the best ways to increase product variety.

Responsive, reliability, resilience, and relationship—the 4Rs are the basis of successful logistic and supply chain management of any business (Christopher 2005, p. 38). Responsiveness gives the ability to a company to respond to customers' demand in shortest possible time. In such fast-changing market place, agility is the key characteristic of a successful company.

Reliability of product could be improved by appropriate control of process and quality of material. Reliability of the logistic service could be improved by increasing the visibility of pipeline. It means that focal company should get more information about far downstream supply chain.

Resilient supply chain enhances the capability of the company to cope with uncertain business environment.

Success of business depends upon the effective management of relationship across the supply chain. Particularly, good relationship between buyer and supplier could reduce cost, improve the quality of work and product, provide better assurance of delivery of raw material/part/sub-assembly/assembly, and prepare a niche for innovative work (Table 3.2)

Postponement integrates aforesaid 4Rs to increase responsiveness of the supply chain. For instance, the use of modular structure of product provides better control of process and quality of material at sub-assembly/component level.

3.6 Drivers of Postponement Strategy

According to Lee (1998), four enablers are identified as follows:

1. Modularity;
2. Design for postponement;
3. Supply chain collaboration; and
4. Associated costs.

Apart from the above, the following prerequisites for postponement are identified as follows:

1. Extensive use of information technology to develop healthy relationship between buyer and supplier;
2. Reduce lead time, particularly after postponement point so that customer waiting time could be reduced as much possible. This will reduce the chances of backorder or lost sales (Lee 1996);
3. Simplification of design and process is required to reduce internal complexity of supply chain; and
4. Postponement is particularly suitable for global supply chain where positioning of inventory, distribution mode, and production facilities are the critical success factors for any organization.

Table 3.2 Categories of postponement strategy

Sl. No.	Type	Definition	Application
1	Pull postponement	In pull postponement or process postponement (Brown et al. 2000), decoupling point moves upstream in the supply chain to make forecast easier (Lee 1998)	1. ABC Bicycle Company, India 2. Benetton, an apparel manufacturer, delays its color dyeing process until orders are received (Lee and Tang 1998)
2	Logistics postponement	Here, customization takes place far downstream (Lee 1998). Packaging postponement and labeling postponement (Twede et al. 2000) or branding postponement (Ackerman 1997) are some of the approaches of logistic postponement	1. Hewlett-Packard produces generic printers at its factory and distributes them to local distribution centers, where auxiliary components are packed 2. All products of IKEA are assembled at customers site
3	Form postponement	Form postponement or product postponement (Brown et al. 2000) opts to combine standardized components with auxiliary components to achieve high customization (Lee 1998 and Brown et al. 2000)	1. Brown et al. (2000) applied form postponement in a semiconductor company (Xilinx) where standard ICs are developed and later they are configured to have customized features as per demand
4	Price postponement	Van Mieghem and Dada (1999) defined price postponement from economic and marketing perspectives. Here, selling price of the product is fixed after receiving customer order	1. Bank of China (BOC) Hong Kong used a price postponement strategy in July 2002
5	Time postponement	Customer order initiates the forward movement of goods	1. Dell computer: Computers are assembled after the orders are placed by customers
6	Place postponement	Goods are distributed after receiving the customer order from a central location	1. Hindustan Unilever Limited, India's largest consumer product company, supplies their products to modern trade segment (organized retail sector), general trade segment and rural markets by using place postponement strategy

3.7 Customer Order Decoupling Point (CODP)

Time-based competition, short product life cycle, volatile market, and heterogeneous customer demand need optimum balance between limited resources of any organization and level of satisfaction of customer to yield desired supply chain surplus. CODP is one of such strategies to balance decision under certainty and uncertainty concerning customer demand (Rudberg and Wikner 2004). According

Table 3.3 Manufacturing strategy and degree of complexity

Type of Strategy	Benefit	Degree of complexity
MTS	Shortest customer waiting time	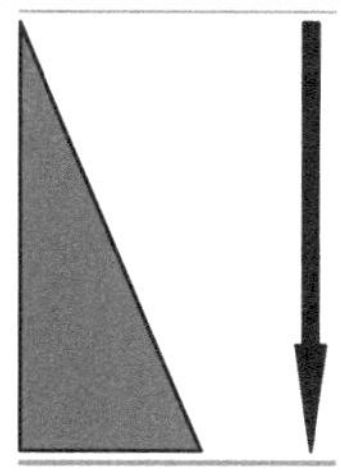
ATO	Reduced WIP inventory, product variety is more than MTS	
MTO	Long customer waiting time comparing to ATO; however, more complex product is manufactured as per the need of customer	
ETO	Long customer waiting time. Particularly suitable for very complex product design ex. Ship manufacturing	

to Olhager (2003), who minted this term, CODP is the point after which customized product is prepared from semi-finished product as per the specification of customer order. He further mentioned that depending upon the position of CODP along the supply chain, different manufacturing strategies can be classified as engineer to order (ETO), make to order (MTO), make to stock (MTS), and assembly to order (ATO) based on the ratio of production lead time and delivery lead time. According to Wikner and Rudberg (2005), mass-customization strategies can be classified into four different categories, as shown in Table 3.3.

3.8 Conclusion

Mass customization is not the panacea for increasing supply chain surplus for any industry. It needs effective strategy, advanced manufacturing technology, advanced information and communication technology, and willingness of customer to pay more for customized product. Customer will pay more for his/her desired product if standard product fails to give his/her desired utility. Companies usually face challenges to manufacture customized product as infinite product variety cannot be developed with their limited resources. Thus, a well-defined product family architecture (PFA) is required. Four mass customization strategies are discussed in this chapter, namely MTS, ATO, BTO, and ETO. Companies such as Dell, Nike, General Motors, and Zara show that with the concept of mass customization they can reduce WIP inventory without effecting overall lead time.

A brief introduction is also included about customer order decoupling point (CODP) and drivers of postponement strategy. CODP is used in supply chain to combine push-type and pull-type strategies. Interested readers are requested to refer relevant journals in this regard to know in detail about CODP and mass customization strategies. Delayed product finalization or postponement is commonly used to reduce manufacture's risk, complexity of supply chain, lead time chain and to enhance the flexibility of supply chain. In this chapter, a new concept of procurement is mentioned in Sect. 3.5, "Sourcing Postponement," to reduce the risk between manufacturer and retailers. Such strategy may reduce WIP inventory, risk for stock out/over stock, and also long waiting time of customer to get his/her product. Sourcing postponement is based on the concept of JIT and useful for ATO industries.

References

Ackerman KB (1997) Postponement, practical handbook of warehousing. Chapman and Hall, New York
Allee V (2000) Reconfiguring the value network. J Bus Strategy 21(4)
Bowersox DJ, Closs DJ (1996) Logistical management: the integrated supply chain process. McGraw-Hill, New York
Brown AO, Lee HL, Petrakian R (2000) Xilinx improves its semiconductor supply chain using product and process postponement. Interfaces 30(4):65–80
Christopher M (2005) Logistics and supply chain management. Prentice Hall Publication
Cheng TCE, Li J, Wan CLJ, Wang S Postponement strategies in supply chain management. Springer 143
Kotha S (1995) Mass customization: implementing the emerging paradigm for competitive advantage. Strateg Manag J 16(Special Issue):21–42
Lee HL (1996) Effective inventory and service management through product and process redesign. Oper Res 44(1):151–159
Lee HL (1998) Postponement for mass customization: satisfying customer demands for tailor-made products. In: Gattorna (ed) Strategic supply chain alignment. Gower, England
Lee HL, Tang CS (1998) Variability reduction through operations reversal. Manage Sci 44 (2):162–172
Olhager J (2003) Strategic positioning of the order penetration point. Int J Prod Econ 85:319–329
Pine II, Joseph B (1993) Mass customization: the new frontier in business competition. Harvard Business School Press, Boston, Massachusetts
Rudberg M, Wikner J (2004) Mass customization in terms of the customer order decoupling point. Prod Plann Control 15(4):445–448
Salvador F, Holan PM, Piller F (2009) Cracking the code of mass customization. MIT Solan Manage Rev 50(3) http://sloanreview.mit.edu/article/cracking-the-code-of-mass-customization/
Tseng MM, Jiao J, Merchant ME (1996) Design for mass customization. CIRP Ann-Manufact Technol 45(1):153–156
Tseng MM, Jiao J (2004) Customizability analysis in design for mass customization. Comput Aided Des. doi:10.1016/j.cad.2003.09.012
Twede D, Clarke RH, Tait JA (2000) Packaging postponement: a global packaging strategy. Packag Technol Sci 13(3):105–115
Van Hoek RI (2000) The thesis of leagility revisited. Int J Agile Manage Syst 2:196–201
Van Hoek RI (2001) The rediscovery of postponement a literature review and directions for research. J Oper Manage 19:161–184
Van Mieghem JA, Dada M (1999) Price versus production postponement: capacity and competition. Manage Sci 45(12):1631–1649
Wikner J, Rudberg M (2005) Integrating production and engineering perspective on the customer order decoupling point. Int J Oper Prod Manage 25(7):623–641
Zipkin P (2001) The limits of mass customization. Solan Manage Rev 42(3):81–87

Chapter 4
Modeling and Optimization of Strategic Sustainable Sourcing

4.1 Introduction

All organizations—but especially large companies and public sector organizations—should use procurement as a way of encouraging those in supply chain to improve environmental performance.

UK Round Table on Sustainable development
Second Annual Report, London, 1997.

The word sustainability is derived from Latin word sustinere. It means to endure or maintain. The concept of sustainability is introduced in the report 'Our Common Future' of the UN commission Brundtland in 1987. According to the definition of sustainable development of the commission Brundtland, sustainable development is the development that meets the needs of the present without spoiling the ability of next generations to achieve their own needs. With the advent of new technologies, man enters the world of luxury and comfort by sacrificing the nature. Unfortunately, ostentatious display of hedonistic society allures people and compels them to forget about divine reality. Present bears the seed of future. Today, we are facing soaring temperature, air pollution, water pollution, etc. because of the selfish activities of mankind. It is crystal clear that creation of wealth should be integrated with eco-ethical lifestyle of human being. Today, companies are augmenting the practice of corporate sustainability and co-creation of value. Traditional supply chain deals with man, money, and material (3M). On the other hand, green supply chain deals with man, money, material, and environment (3Me). Finally, sustainable supply chain deals with man, money, material, environment, and society (3MeS).

K. Mukherjee, *Supplier Selection*, Studies in Systems, Decision and Control 88,
DOI 10.1007/978-81-322-3700-6_4

4.2 Literature Review

Do procurement policies need to change and if so how? What sort of procurement policies are required to achieve sustainability or green procurement status? To find the answers of aforesaid questions, author searched thoroughly different literature. Literature review shows that before 1980s, purchasing function got scant attention from corporate sector. It was considered as simple clerical job. Later on, people understood the very essence of procurement and tried their best to use it as mere tool to reduce overall cost of procurement. Such myopic view compelled several academicians to explore hidden benefits of procurements process. At the end of 1980s, clean technologies were introduced for reducing the environmental impact. At the beginning of 1990s, enterprises changed their operating procedures to introduce eco-auditing frameworks to rectify products and services (Franke 1995). Recently, environmental conscious firms, mainly large firms are organizing environmental programs to alter their supply chain (Gupta 1995). For instance, Sony's procurement policy promotes 'QCDS + E', i.e., quality, cost, delivery, service, and environment (Handfield et al. 2002). Brink et al. (1998) developed a system called 'ECO-QUEST' to prepare questionnaire for suppliers to self-audit their environmental performance and gives them some solutions to improve environmental profile of their product. Noci (1997) suggested a preliminary framework that identifies measures for assessing environmental performance but gave little attention on environmental cost data. Because of such cumulative efforts, today's purchasing managers are working as strategic sourcing decision makers to offer better products at lower cost, better quality, and with greater flexibility. Strategically managed long-term relationships with key suppliers have a positive impact on the firm's financial performance and intensity of buyer–supplier coordination (Carr and Pearson 1999; De Toni and Nassimbeni 1999). Managers and policy makers are gradually realizing that purchasing can redo corporate function/performance along environmental dimension (Handfield et al. 2002). Procurement of any company can affect environment directly and indirectly (Handfield et al. 2002). Direct effect of product is the wastage of product during transportation, storage, processing, use, or disposal. When a company is purchasing the product, it is indirectly integrating environmental effect of product with its value chain. Awareness of green procurement is germinated from several new happenings, such as the follows:

1. Introduction of ISO 14000 certification standard which focuses the firm's environmental management system (EMS). Procurement is one of the key assets of **ISO 14000** to control disposal as well as procurement of material.
2. Pressure of government agencies like **Environmental Protection Agencies (EPA)** to control waste reduction.

Dickson (1966) in his seminal work proposed 23 criteria for supplier selection. Pareto analysis is further conducted to find most cited criteria, as shown in Fig. 4.1.

Among 23 criteria, some criteria such as performance history, reputation and position in history, impression and amount of past business are very closely related.

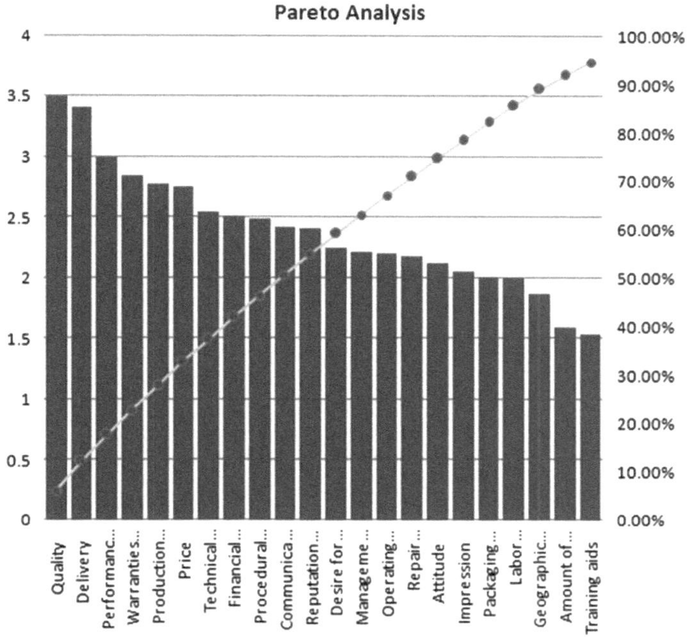

Fig. 4.1 Pareto analysis of Dickson's 23 criteria

Higher amount of past business shows companies market share as well as its reputation. On the other hand, amount of past business is a part of performance history. However, Dickson's 23 criteria gave a ready-made platform for academician as well as industrial practitioners. Out of the 23 criteria, only packaging ability, warranties and claim policies, and technical capabilities are closely related to green supply chain. Moreover, Dickson gave an indication of sustainability through labor relations record which is the social responsibility of supplier. Question arises about the viability of Dickson's 23 criteria for green supplier selection. Can we use the same 23 criteria for green procurement process? To find the answer of this question, author conducted a survey of recent literature, as shown in Table 4.1. Methods such as AHP, CBR, DELPHI, DEMATEL, gray entropy, etc. are used to trade off palpable and non-palpable criteria of supplier selection. Majority of the research works on green supplier selection are based on automotive industry, apparel industry, electronic industry, or closely related sectors. It is clear that green procurement is not a mere optimization or decision-making problem. It is a strategy to integrate mission and vision of any company to broaden the way of sustainability.

Humrhreys (2003) proposed a hierarchical model of supplier selection for reversed supply chain. A brief analysis of his model with critical remarks is shown in Table 4.2.

Table 4.1 Literature review of green supplier selection

Sl. No.	Author	Year	Criteria	Method
1	Robert Handfield, Steven V. Walton, Robert Sroufe, Steven A. Melnyk	2002	Product attributes, waste management, labeling/certification, packaging/reverse logistics, compliance to government regulations, environmental programs at the supplier facilities.	AHP, Delphi
2	Paul Humphreys, Ronan McIvor, Felix Chan	2003	Environmental competencies, environmental management system, design for environment, green image, management competencies, environmental costs	CBR and multi-attribute analysis
3	P.K. Humphreys, Y.K. Wong, F.T.S. Chan	2003	—do—	Multi-attribute analysis
4	YANG Yuzhong, WU Liyun	2007	Quality, price, capability, service, environment protection, management system, and credit standing	Grey entropy
5	Satish Nukala and Surendra M. Gupta	2007	Quality, on-time delivery, proximity, cultural and strategic issues	Taguchi loss function, AHP, fuzzy.
6	C. Y. Chiou, C. W. Hsu, W. Y. Hwang	2008	Supply chain base: Cost, quality, delivery, service performance, Green Competencies: green material selection, cleaner production technologies, reduced green packaging, reverse logistics, Environmental management systems: ISO-14001 certification, Eco-labeling, supplier environmental evaluation, environmental management information system, Environmental performance: green design, use of toxic/restricted substances, Corporate social responsibility: Public disclosure of environmental record, green image, relationship with stakeholders, Contributions to community, Risk Factor: Government and local rules/regulations, Political stability, Supplier's credit, Internal management system	Fuzzy analytic hierarchy process
7	YU-ZHONG YANG, LI-YUN WU	2008	Quality, price, capability, service, and environment protection, management system and credit standing	Extensible synthetic evaluation
8	Amy H.I. Lee, He-Yau Kang, Chang-Fu Hsu, Hsiao-Chu Hung	2009	Quality: quality-related certificates, capability of quality management, capability of handling abnormal quality, Finance: Past finance performance, stability of finance, Price, Organization: attitude of managers, future strategy direction, degree of strategic cooperation,	Fuzzy set theory, AHP

(continued)

Table 4.1 (continued)

Sl. No.	Author	Year	Criteria	Method
			Technology capability: capacity, technology level, capability of R&D, capability of design, capability of preventing pollution, Service: credible delivery, capability of delivery on time, capability of technology support, flexibility, Total product life cycle cost: cost of supplied components, Green Image: green purchase trend of customer, Pollution control: Use of harmful materials. Environment management: Environment-related certificates, internal control process	
9	GE Yan	2009	Environment, quality, operational capacity, price level, service level,	AHP and GA
10	Cao Qingkui, Ruan Junhu	2009	Product information: delivery time, price, quality, Geographical position: delivery location, delivery way, Green information: resource utilization ratio, Resource recovery ratio, resource types, Resource conservation, environmental pollution, environmental governance, Cooperative history: cooperation duration, cooperation condition, Service: Response time, service level	Rough sets and AHP
11	Mehrdad Agha Mohammad Ali Kermani, Amir Malaei, Marzieh Nasiri	—	Price, quality, services, and environmental criteria	AHP and multi-objective optimization (presented mathematical model for sustainability)
12	Alireza Iirajpour, Mehdi Hajimirza, Mahdi Golsefid Alavi, Sajad Kazemi	2012	Design management—process/product changeability, ability to design and recycle Technology management—technology level, ability of R&D, clean technology, Customer management—controlled use of hazardous material, energy consumption, green packaging Strategic management—green image, green planning, supervision on and following environmental laws, social responsibility	DEMATEL

Table 4.2 Critical remarks for Humrhreys model for reversed supply chain

			Critical remarks
Quantitative cost	Pollutant costs/effect	Solid waste	Saving energy means saving money. Thus, energy should be recovered from used product. However, care should be taken to reduce exhaust to air, water, and soil as much as possible
		Chemical waste	
		Air emission	
		Water waste disposal	
		Energy	Optimum use of energy should be considered from the procurement of raw material to shipment to end users
	Improvement cost	Buying environmental free material	Select material as per product specification to reduce environmental cost. In this regard, author suggests using SimaPRO or Gabi 4 software to calculate the energy associated with each raw material
		Buying new environmental free technology	Earlier trend of manufacturing gave highest priority to low cost. Environment consciousness gives highest priority to cost and environmental burden associated with the manufacturing/assembling of product
		Redesign of product	It is basically amalgamation of design for assembly, design for disassembly and environmental conscious manufacturing process
		Staff training	It is required to make employee environment conscious
		Recycling	Every recycling is associated with profit. To maximize the profit, company should reuse the complete product or partly disassemble the product as the cost of disassembly initially decreases and then increases with the degree of disassembly. Therefore, author strongly suggests that each product should be associated with optimum disassembly plan
	Management competencies	Senior management support	
		Environment partners	
		Training	Training gives required competencies to employee. It also creates an awareness of eco-friendly product
		Information exchange	Reversed supply chain is very complex in nature. Like forward supply chain, sharing of information is also required for reverse supply chain
	Green image	Customer's purchasing retention	Green market share will increase if customer is willing to purchase green/eco-friendly product. Government policies also play a major role to control green market share of any product
		Green market share	
		Stakeholders relationship	
	Design for environment	Recycle	Complexity of reverse supply chain is more than forward supply chain. However, effective strategies for reverse supply chain can reduce cost and environmental burden associated with product
		Re-use	
		Remanufacturing	
		Disassembly	
		Disposal	
	Environmental management systems	Environmental policies	Such factors are considered to measure the capability of supplier to supply eco-friendly product/parts/sub-assembly/raw material to the customer.
		Environmental planning	
		Implement and operation	
		ISO 14001 certification	

(continued)

Table 4.2 (continued)

			Critical remarks
	Environmental competencies	Clean technology availability	Environmental management systems can be considered as cause and environmental competencies can be considered as its effect
		Use of environment friendly materials	
		Pollution reduction capability	
		Returns handling capability	

4.2.1 *Viability of Dickson's 23 Criteria for Green Supply Chain*

Dickson's 23 criteria have been considered as ready-made reference for forward supply chain. However, author suggests that Dickson's criteria can be used for reverse supply chain with little modification. Table 4.3 shows the list of few unchanged, changed, and new entrant criteria.

4.3 Economical Aspects of Reverse Supply Chain

Does reverse supply chain pay? One of the main advantages of reverse supply chain is to enhance supply chain surplus by mixing reused parts/products with new product. It saves time of manufacturing, reduce total cost of procurement but

Table 4.3 New criteria for green supplier selection

Sl. No.	Dickson's criteria	Green supplier selection criteria
Changed criteria		
1	Management and organization	Environmental management system, management competencies
2	Training aids	Training aids
3	Desire for business	Green image
4	Technical capability	Ability of R&D, Cleaner production technology
5	Packaging ability	Green packaging ability
6	Procedural compliance	Compliance to government regulations for eco product
New criteria		
1	New entry	Corporate social responsibility
2	New entry	Design for environment
Unchanged criteria		
1	Quality	Quality
2	Delivery	Delivery
3	Price	Price

enhances the complexity of supply chain. It also raises questions of quality to reuse dumped parts/products. Thus, an optimal balance is required.

4.3.1 Disassembly Cost

A disposed product is composed of several parts, each of which may have some material value or reuse. The purpose of the disassembly operation is to retrieve valuable parts and then forward those to the appropriate recycling streams (Das and Yedlarajiah 2002). Output of the disassembly operation could be waste material, recyclable material, or parts for direct use. In general assembled product is disassembled either manually or through automatic process. For each customized product, a suitable disassembly plan should be developed to disassemble the product with less time to reduce the disassemble cost. In this regard, various researchers have developed various methods to calculate disassembly cost. Special emphasize has been given to the design of fasteners. Ishii et al. (1994) proposed the disassembly cost DC as

$$\mathrm{DC} = \sum_{i=1}^{l} C_i + \sum_{j=0}^{m} (\mathrm{fn} * \mathrm{F})_j + \sum_{k=0}^{n} (\mathrm{pn} * \mathrm{p})_k$$

C_i time to remove component
F_J time to remove fastener
f_{nJ} number of fastener associated with one link
p_{nk} number of process points
p_k time to remove or undue process
l total number of components in the system
m total number of links with fastener
n total number of links with fastening process.

The calculated disassembly time associated with each product should be multiplied with labor wages per unit time. Obvious question arises about the importance of disassembly process in environmental conscious manufacturing. In this regard, Das and Mathew (1999) found that there are eight common class of material output bins that are maintained in electronics disassembly facilities; these are ferrous metals (steel), non-ferrous metals (aluminum, copper), sources of precious metals (gold, silver, palladium), packaging materials (styrofoam), glass, ceramics, plastics, hazardous parts, and paper. Hence by integrating disassembly process, companies could reduce their production cost and thereby maximize their profit. Das et al. (2000) propose a scheme for the estimation of **disassembly effort index (DEI)** which is the function of seven factors—(a) time (b) tools (c) fixture (d) access (e) instruct (f) hazard, and (g) force required to disassemble a product. The *DEI* scores ranges from zero upward, with zero indicating no effects. Thus, disassembly is a crucial factor in implementing environmental conscious manufacturing (Zeid and Gupta 2002).

4.3.2 *Recycling Profit*

A product recycling module should be developed in this regard to calculate the recycling profit of the disposed product. Zhang et al. (2004) calculated the profit due to recycling as PR

$$\mathrm{PR} = \mathrm{qp} * \mathrm{wt} - (\mathrm{dc} + \mathrm{ic} + \mathrm{rc} + \mathrm{tc})$$

where

qp quote price in the market
wt weight of the material recovered
dc disassembly cost
ic inspection cost
rc recycling cost
tc transportation cost.

The product recycling module will interact with LCA module and depending upon the post consumer use of the product (i.e., reuse, recycling, incineration and landfill), LCA calculation will be modified.

4.3.3 *Optimum Level of Disassembly*

Complete disassembly is not profitable as the marginal cost benefit for complete disassembly becomes less attractive (Das and Naik 2002; Gungor and Gupta 1999). Total disassembly cost increases with the level of disassembly. The lowest total disassembly cost determines the termination of the disassembly process (Zhang et al. 2004). A proper disassembly process planning (DPP) is required to disassemble the product up to a certain level to reduce the cost of disassembly (Gungor and Gupta 1999). Researchers defined the cost of disassembly in several ways. For instance, disassembly cost is composed of direct labor cost, disassembly effort cost, and the logistic cost (Das and Naik 2002). According to Zhang et al. (2004), total cost of disassembly is composed of disassembly cost which includes labor, and tooling cost; material reprocessing cost etc.

4.4 Sustainable Mass Customized System

Author suggests a new way of mass customization to encompass the concept of sustainability with the concept of mass customization. Sustainable mass customized system interacts with planet (the environment) to reduce the emission of green house gases (GHG), people (the customer as well employee) to maintain a healthy social life, product with customizable characteristics to give the desired utility to individual customer and prosperity (the economic growth) to the organization. Therefore, sustainable mass customized system can be considered as a system of people, planet, product, and their combined prosperity.

4.4.1 Stages and Enablers of Sustainable Mass Customized System

Green procurement, environmental conscious manufacturing (ECM), green product design (GPD), reverse logistics, reuse and recycle, etc. are commonly used to make green supply chain. It is mentioned in the bottom layer of triangle in Fig. 4.2. For instance, company should look after the welfare of employee's family. Such policies enhance the integrity of the employee with his company. It is mentioned in middle layer of triangle. After achieving the sustainable supply chain, company should proceed for auditing and bench marking. It is the top layer of sustainable triangle. Considering the need of sustainable supply chain, the author suggests that mass customization strategies should incorporate sustainable issues to make profitable, eco-friendly, and eco-ethical ambiance for the organization.

4.4.2 Sourcing Strategies for Sustainable Mass Customized System

Any complex product can be developed through modular approach. Usually, companies diversified their portfolio of product to manufacturer different brand to enhance profit margin through economy of scopes. For instance, each and everyday

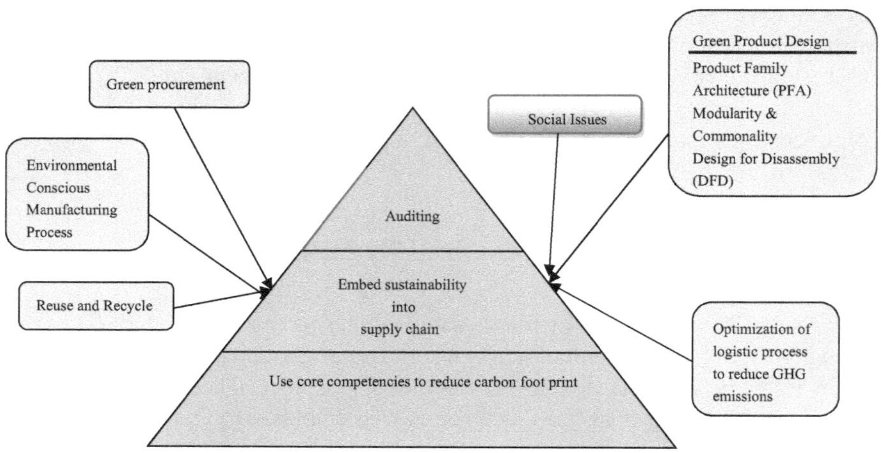

Fig. 4.2 Six enablers of 3-stage sustainable mass customized system

users are receiving cell phone with different varieties. It is reducing product life cycle, reducing time to introduce new product in market, and at the same time triggering tough competition among competitors. To survive in such situation, companies prefer to develop modular product. A modular product composed of several standard parts/components. Standard parts/components can be manufactured as per the aggregated demand and auxiliary parts can be developed and assembled with standard or base parts on receipt of the customer's request. This approach can reduce the complexity of inventory, reduce the chances of over stock or under stock, and enhances the responsiveness of supply chain. It is an integrated approach of push and pull strategy of supply chain. Therefore, base product can be made with aggregated demand and auxiliary product can be made with stochastic demand. This 2-stage procurement problem can be solved with intuitionistic fuzzy AHP (IF-AHP) and multi-objective genetic algorithm (MOGA).

4.5 Mathematical Model for SPS

China has emerged as a favored low-cost manufacturing spot among electronics companies because of the presence of competent suppliers and contract electronics manufacturers, as well as the quality of its infrastructure (roads, electricity, stable political situation, etc.). Although electronics assembly may take only one to two days, transporting goods by ship between China and Europe takes three weeks. Such delay in service can create several problems. One solution to this problem is to create standard products in the low-cost production center but does final configuration and packaging at a distribution point closer to the customer. Algorithm of this two-stage sustainable procurement process is mentioned below:

Stage-1: Selection of suppliers for base product with aggregated demand

1. Prepare intuitionistic fuzzy pairwise comparison matrix for each criterion and alternative.
2. Calculate score (S_i) of all intuitionistic fuzzy number with any of the given formula.

$$\mathrm{S_I}(X_{ij}) = \mu_{ij} - v_{ij} \text{ where } \mathrm{S_I}(X_{ij}) \in [-1, 1] \tag{4.1}$$

$$\mathrm{S_{II}}(X_{ij}) = \mu_{ij} - v_{ij} \cdot \pi_{ij} \text{ where } \mathrm{S_{II}}(X_{ij}) \in [-1, 1] \tag{4.2}$$

$$\mathrm{S_{III}}(X_{ij}) = \mu_{ij} - (v_{ij} + \pi_{ij})/2 \text{ where } \mathrm{S_{III}}(X_{ij}) \in [-0.5, 1] \tag{4.3}$$

   $\mathrm{S_I}(\mathrm{X_{ij}})$ is useful for simple decision-making problem, and $\mathrm{S_{II}}(X_{ij})$ and $\mathrm{S_{III}}(X_{ij})$ are useful for complex decision-making problem (Chen 2011).
3. Calculate normalized score matrix with the given formula

$$\overline{s_{ij}} = \frac{s_{ij} - \min\limits_{j}\{s_{ij}\}}{\max\limits_{j}\{s_{ij}\} - \min\limits_{j}\{s_{ij}\}} \tag{4.4}$$

4. Normalize each row of $\overline{S}$ with the given formula

$$\dot{\overline{s}} = \frac{\overline{s_{ij}}}{\sum_{j=1}^{n} \overline{s_{ij}}} \forall i = 1, 2, 3 \ldots m \text{ and } j = 1, 2, 3 \ldots n \tag{4.5}$$

5. Calculate entropy w.r.t ith attribute with the given formula

$$E_i = -\frac{1}{\ln n} \sum_{j=1}^{n} \dot{\overline{s_{ij}}}\, ln\, \dot{\overline{s_{ij}}} \tag{4.6}$$

6. Calculate entropy weight w_i with the given formula

$$w_i = \frac{1 - E_i}{\sum_{i=1}^{m} (1 - E_i)} \tag{4.7}$$

7. Calculate normalized entropy weight to rank criteria or alternative with the given formula

$$\overline{w_i} = \frac{W_i}{\sum_{i=1}^{m} W_i} \tag{4.8}$$

8. Prepare weighted order allocation model with the priority obtained from step-7. The following notations are used to prepare multi-objective order allocation model mentioned below:

C_{ij}	Purchase cost of product j from ith supplier
TC_{ij}	Transportation cost of product j from ith supplier
CC_i	Overall performance index of ith supplier
C_{Oij}	Ordering cost of jth product from ith supplier
α_i	Reliability of ith supplier
X_{ij}	Order quantity of product j to ith supplier
LD_{ij}	Percentage late delivery of product j from ith supplier
V_{ij}	Capacity of ith supplier for jth product
D_j	Demand for jth product
H_j	Handling cost per ton of product j
B	Total allocated budget for all products
λ_j	Percentage of jth product disposed at disposal site
ξ_j	Level of disassembly of jth product at disassembly site

β_i GHG emission factor per weight unit distance due to use of transportation mode
d_i Distance of ith supplier from manufacturing/retailing site
alpha Probability value of chance constraint
i 1,2,3…n of suppliers
j 1,2,3…m no of products

Total cost of purchase (TCP) consists of purchase, transportation, order/setup, and holding costs.
Min TCP:

$$\sum_{i=1}^{n}\sum_{j=1}^{m} C_{ij}X_{ij} + \sum_{i=1}^{n}\sum_{j=1}^{m} \mathrm{TC}_{ij}X_{ij} + \sum_{j=1}^{m} H_j \sum_{i=1}^{n} X_i + \sum_{i=1}^{n}\sum_{j=1}^{m} C_{Oij}X_{ij} \tag{4.9}$$

Maximize TVRP:

$$\sum_{i=1}^{n}\sum_{j=1}^{m} \alpha_i \mathrm{CC_i} X_{ij} \tag{4.10}$$

Minimize number of late deliveries:

$$\sum_{i=1}^{n}\sum_{j=1}^{m} \mathrm{LD}_{ij}X_{ij} \tag{4.11}$$

Minimize GHG emission for inbound logistics:

$$\sum_{i=1}^{n} \beta_i d_i \sum_{j=1}^{m} X_{ij} \tag{4.12}$$

Subject to
Capacity constraint:

$$\sum_{j=1}^{m}\sum_{i=1}^{n} X_{ij} \le \sum_{j=1}^{m}\sum_{i=1}^{n} V_{ij} \quad \text{for} \quad i = 1, 2, 3\ldots n \quad \text{and} \quad j = 1, 2, 3\ldots m \tag{4.13}$$

Demand constraint:

$$\sum_{j=1}^{m}\sum_{i=1}^{n} X_{ij} = \sum_{j=1}^{m} D_j - \sum_{j=1}^{m} (1-\lambda_j)\xi_j D_j \tag{4.14}$$

Cost constraint:

$$\sum_{j=1}^{m}\sum_{i=1}^{n} C_{ij}X_{ij} \leq B \tag{4.15}$$

Non-negativity constraint:

$$X_{ij} \geq 0 \text{ for } i = 1, 2, 3 \ldots n \text{ and } j = 1, 2, 3 \ldots m. \tag{4.16}$$

Stage-II: Selection of suppliers for auxiliary product with stochastic demand

9. Repeat steps 1–7 mentioned in stage-1 to find the priority of each supplier and form the weighted order allocation model mentioned below:
   Minimize TCP:

$$\sum_{i=1}^{n}\sum_{j=1}^{m} C_{ij}X_{ij} + \sum_{j=1}^{m} H_j \sum_{i=1}^{n} X_i + \sum_{i=1}^{n}\sum_{j=1}^{m} C_{Oij}X_{ij} \tag{4.17}$$

Maximize TVRP:

$$\sum_{i=1}^{n}\sum_{j=1}^{m} \alpha_i \mathrm{CC_i} X_{ij} \tag{4.18}$$

Minimize number of late deliveries:

$$\sum_{i=1}^{n}\sum_{j=1}^{m} \mathrm{LD}_{ij}X_{ij} \tag{4.19}$$

Subject to

10. Demand constraint:

$$\mathrm{pr}\left[\sum\nolimits_{i=1}^{n} x_{ij} \geq D_j\right] \geq \text{alpha} \tag{4.20}$$

Above chance constraint can be converted to a deterministic constraint as follows (Guan et al. 2007):

11. $$\sum_{i=1}^{n}\sum_{j=1}^{m} X_{ij} \geq \mu_j + \Phi^{-1}(\text{alpha})\sigma_j \tag{4.21}$$

12. where

$$\Phi(X) = \frac{1}{\sqrt{2\pi}\sigma} \int_{-\infty}^{X} e^{-\frac{(X-\mu)^2}{2\sigma^2}} dX = \text{alpha}, \ \Phi^{-1}(\text{alpha}) = X \quad (4.22)$$

13. Capacity constraint:

$$\sum_{j=1}^{m} \sum_{i=1}^{n} X_{ij} \leq \sum_{j=1}^{m} \sum_{i=1}^{n} V_{ij} \quad (4.23)$$

14. Cost constraint:

$$\sum_{j=1}^{m} \sum_{i=1}^{n} C_{ij} X_{ij} \leq B \quad (4.24)$$

15. Non-negativity constraint:

$$X_{ij} \geq 0 \text{ for } i = 1, 2, 3 \ldots n \text{ and } j = 1, 2, 3 \ldots m. \quad (4.25)$$

4.6 Decision Support System for Strategic Sustainable Sourcing in Volume Discount Environment

In the late 1970s, decision support tools were developed in DOS and UNIX environment. In early 1990s, it was developed in Windows environment. With the development of data processing, microprocessor and networking technology DSS tools became more flexible, user friendly, and be able to solve complex decision-making problem with speed and accuracy. The development of wireless network and cell phones made next big leap of DSS tools. DSS used for supplier selection commonly uses a database, a knowledge base, a graphical user interface, a optimization tool, and other analytical tools to make decision-making process more flexible as well as presentable. Research on sustainable supplier selection is still in infancy. Only the work of Amindoust et al. (2012) has been identified on fuzzy inference system for sustainable supplier selection. They used total 132 rules to select sustainable suppliers. They also claimed robustness and flexibility of their proposed model. But robustness of their model may not true as defuzzification value changes for all membership functions except triangular membership functions. Robustness of the proposed model of Amindoust et al. (2012) is true as long as users are using triangular fuzzy membership function only. Secondly, they did not tried to reduce the rule base of their proposed system. For instance, at stage 1 rule base could be reduced from 16 to only 4 with SVD method. The complicated output surface at stage 1 of their proposed model can also be simplified as shown in

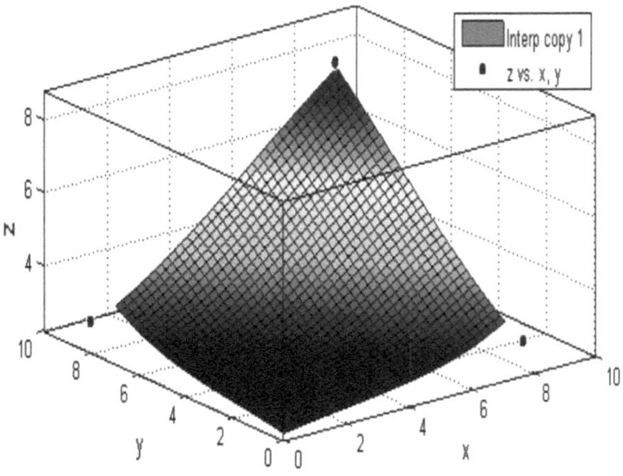

Fig. 4.3 SVD simulation using four rules at stage 1

Fig. 4.3. Membership function at the input can be modified to two odd-shaped membership functions as shown in Fig. 4.4.

Finally, Amindoust et al. (2012) could enhance the flexibility of their model by integrating it with any stand-alone or online software to evaluate a large number of suppliers. In this section, detailed discussion is provided to design and develop decision support system with FIS. The proposed FIS is developed with **intersection rule configuration (IRC)** model for **mamdani fuzzy system** for 2-input and 1-output at each stage. **Triple bottom line** approach is used for proposed model, as shown in Fig. 4.5.

Price, quality, service, delivery, capacity, and past performance are selected for economic criteria. Environmental competencies, environmental management system, green packaging ability, and green product design are selected for environmental criteria. Health and safety, and education are selected for social criteria. All sub-criteria are considered to be larger-is-better. Here, term price refers the difference between the highest quoted price and the price quoted by *i*th supplier. Thus, higher the price means lower the price quoted by *i*th supplier. Triangular membership functions, shown in Figs. 4.6 and 4.7, are selected for input and output in stage 1. Membership functions shown in the figure are selected for output in stage 2 and for input and output in stage 3. Mamdani fuzzy inference system consists of four components—fuzzifier, fuzzy rule base, defuzzifier, and interface engine. Fuzzifier contains different membership functions, such as triangular fuzzy membership function, trapezoidal

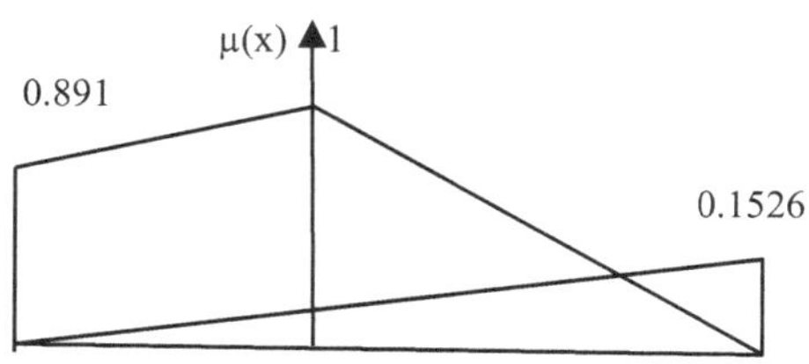

Fig. 4.4 Two odd-shaped membership functions for input at stage 1

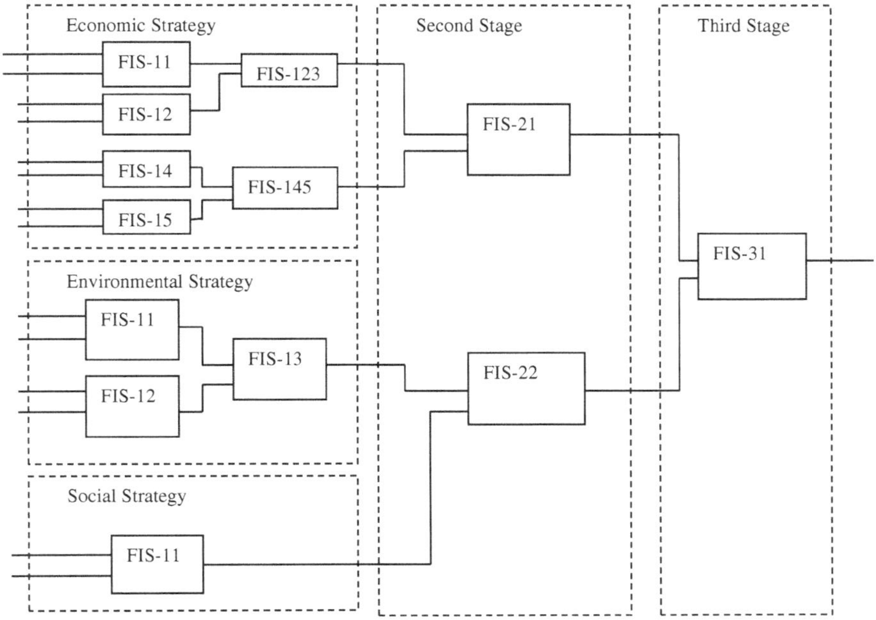

Fig. 4.5 FIS model for sustainable supplier selection

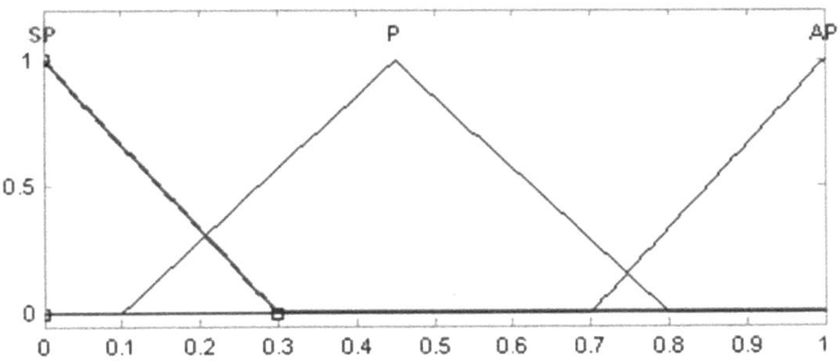

Fig. 4.6 Membership function for stage 1

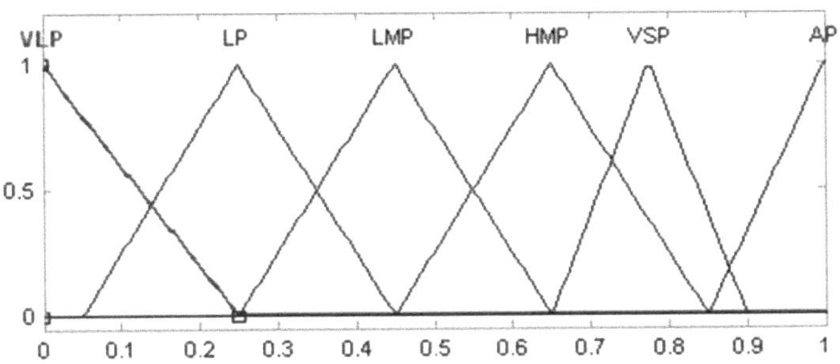

Fig. 4.7 Membership function for stages 2 and 3

Table 4.4 Linguistic terms in stage 1 and stage 2

Slightly preferred (SP)	(0, 0, 0.3)
Preferred (P)	(0.1, 0.45, 0.8)
Absolutely preferred (AP)	(0.7, 1, 1)

Table 4.5 Linguistic terms in stage 2 and stage 3

Very less preferred (VLP)	(0, 0, 0.25)
Less preferred (LP)	(0.05, 0.25, 0.45)
Low moderately preferred (LMP)	(0.25, 0.45, 0.65)
High moderately preferred (HMP)	(0.45, 0.65, 0.85)
Very strongly preferred (VSP)	(0.65, 0.775, 0.9)
Absolutely preferred (AP)	(0.85, 1, 1)

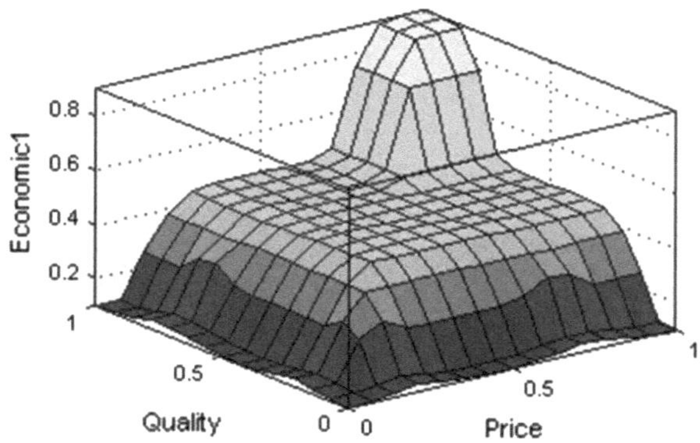

Fig. 4.8 Output surface of FIS for quality and price

membership function, and Gaussian membership function. Defuzzifier includes center of area method (COA) and bisector of area method (BOA). The linguistic terms used for proposed FIS are shown in Tables 4.4 and 4.5.

As shown in Fig. 4.8, economic strategy 1 sharply increases with the increase in quality and price. It means supplier who will offer low cost and better quality is expected to score well. As shown in Fig. 4.9, overall economic strategy increases with the increase in economic strategy 1 and 2. It indicates that a supplier can expect to score well if he offers low cost, better quality, better service, and delivery. For instance, if a supplier pays more priority to economic strategy 1&2 and less to economic strategy 3, then he can expect to score more, as shown in Fig. 4.10. This is one of the flexible options of the proposed FIS as the rule base can be prepared as per the long-term goal of the company. A simulink model is prepared with constant input, as shown in Fig. 4.11, to connect all FIS models. Similarly other FIS models are prepared for environmental and social strategies. As shown in Fig. 4.12, overall increase obtained from combined strategy is observed for an increase in environmental and social strategies. As shown in Fig. 4.13, a supplier should pay same

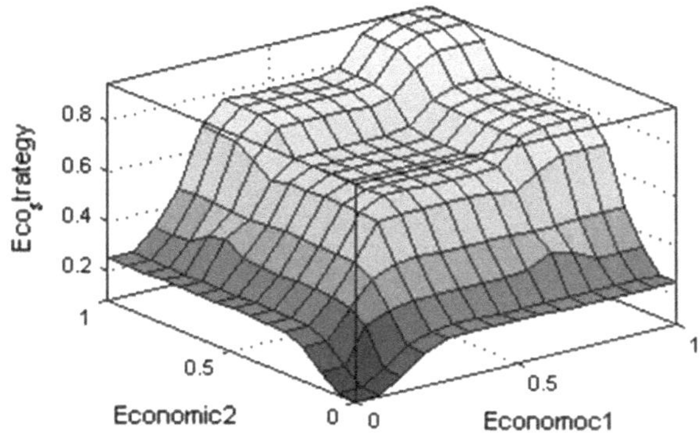

Fig. 4.9 Output surface of FIS for economic strategies 1 and 2

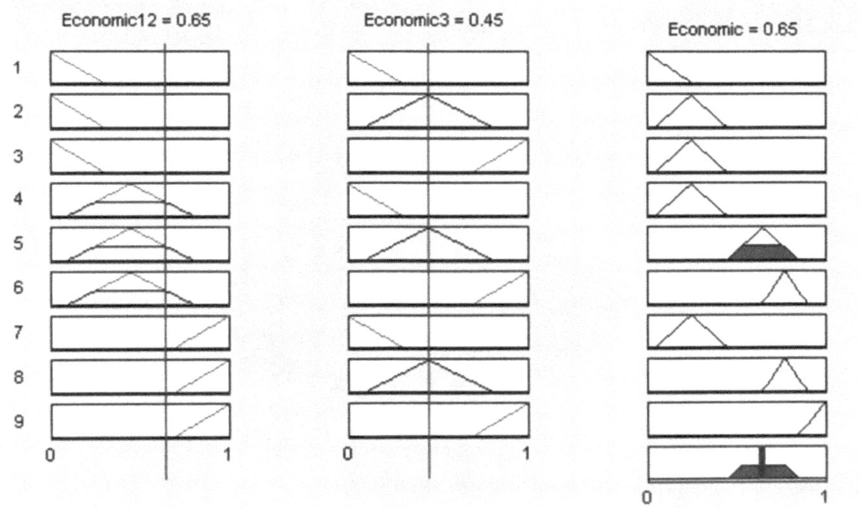

Fig. 4.10 Rules for overall economic strategy

priority to both environmental and social criteria to score well in environmental and social strategy. Thus, the proposed FIS shows that low cost, better quality, or better service is not the market-winning strategy for any supplier to get order. In sustainable procurement process, supplier has to pay more attention on overall improvement of his performance for each of social, environmental, and economic criteria. It is crystal clear that if a supplier pays more attention on economic criteria and pays less on environmental and social criteria, then he can expect to get poor score, as shown in Fig. 4.14. In Fig. 4.15, detailed simulink model is shown for proposed cascaded FIS model.

The proposed FIS model is capable to select large number of suppliers. The proposed model is integrated with VB.NET and SQL server. A simple website is

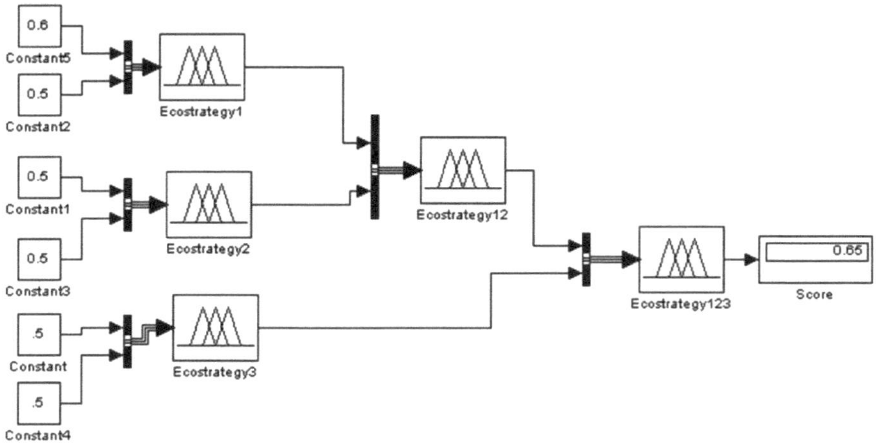

Fig. 4.11 Simulink model for economic strategy

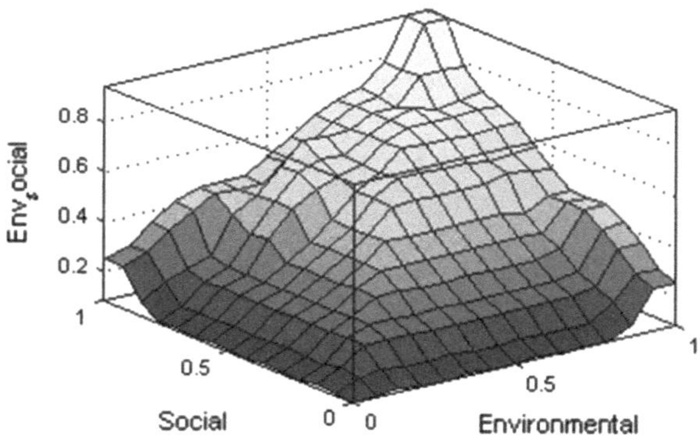

Fig. 4.12 Rule surface for environmental and social strategy

developed to collect potential suppliers' details to prepare supply base for future requirements, as shown in Fig. 4.16. Proposed FIS model remains on a standalone computer. From website, the details of suppliers are stored in .xls file to feed the simulink model.

Following notations are used to prepare order allocation model as shown below:

P_i Purchase cost of per ton of raw material from *i*th supplier
CC_i Rank of *i*th supplier
α_i Reliability of *i*th supplier
X_i Order quantity to *i*th supplier
V_i Capacity of the *i*th supplier
D_j Demand of *j*th product in a planning horizon
r_i Volume discount given by *i*th supplier

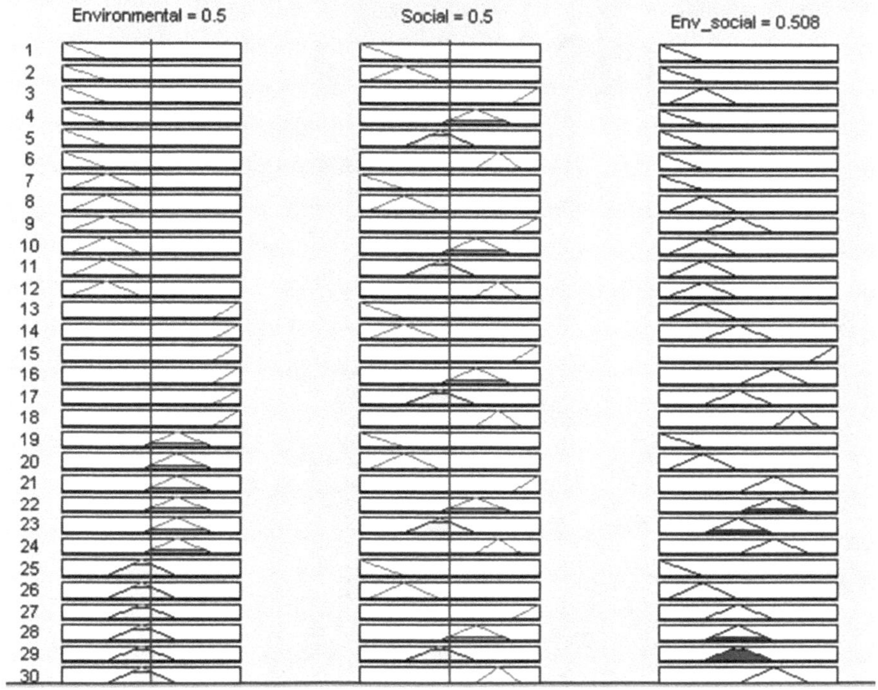

Fig. 4.13 36 rules for environmental and social strategy

Minimize total cost of purchase (TCP): $\sum_{i=1}^{n} (1 - r_i)P_iX_i$
where r_i is defined as follows:

$$r_i = \begin{cases} 0 & \text{if } 0 \le P_iX_i < 10,000 \\ 10 & \text{if } 10000 \le P_iX_i < 20,000 \\ 20 & \text{if } P_iX_i \ge 20000 \end{cases}$$

The above *ri* is based on the model of (Xia and Wu 2007).
Maximize the total value of reliable purchase (TVRP): $\sum_{i=1}^{n} \alpha_i \mathrm{CC_i} X_i$
Subject to
Capacity constraint: $X_i \le V_i \; \forall i = 1, 2, 3\ldots n$
Demand constraint: $\sum_{i=1}^{n} X_i = D_j \; \forall i = 1, 2, \ldots n$ and $j = 1, 2, \ldots m$
Cost constraint: $\sum_{i=1}^{n} (1 - r_i)\, P_iX_i \le 3, 000, 000$
Non-negativity constraint: $X_i \ge 0 \; \forall i = 1, 2, 3\ldots n$
Each supplier is giving three volume discounts. Three suppliers will give 27 (3^3) different combinations of equations (Xia and Wu 2007).

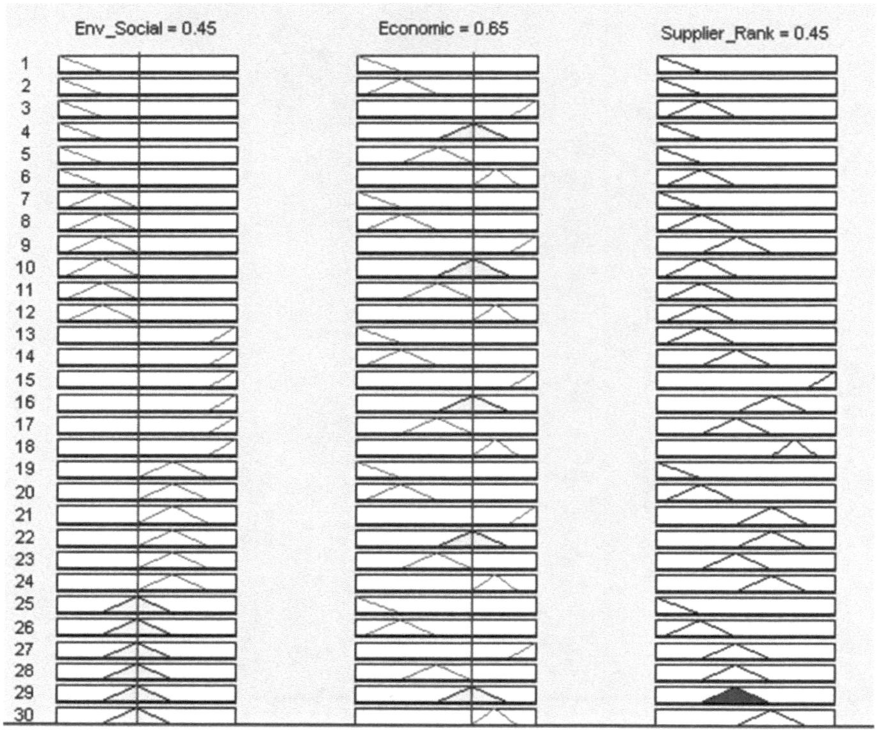

Fig. 4.14 Rules for economic, social, and environmental strategies

Example

A company is willing to initiate sustainable procurement process and going to select suppliers for its product. Volume discount offered by each supplier is shown in Table 4.6. For sake of simplicity, it is considered that all suppliers are offering same volume discount. Supplier datasheet is shown in Table 4.7. Demand of the product in planning horizon is 1200 ton, and total budget is Rs 3 crore. An appropriate method should be used to select sustainable suppliers.

Proposed cascaded FIS is used to rank suppliers. Only three best suppliers are selected. Following objective functions and constraints are prepared as per the supplier datasheet, as shown in Table 4.7.

Minimize total cost of purchase (TCP): $(1 - r_1) * 1500 * X_1 + (1 - r_2) * 1499 * X_2 + (1 - r_3) * 1501 * X_3$

Maximize total value of reliable purchase (TVRP): $0.95 * 0.5 * X_1 + 0.95 * 0.45 * X_2 + 0.96 * 0.508 * X_3$

Demand constraint: $x1 + x2 + x3 = 12,000$

Capacity constraint: $x1 \leq 5000; x2 \leq 6000; x3 \leq 6000$

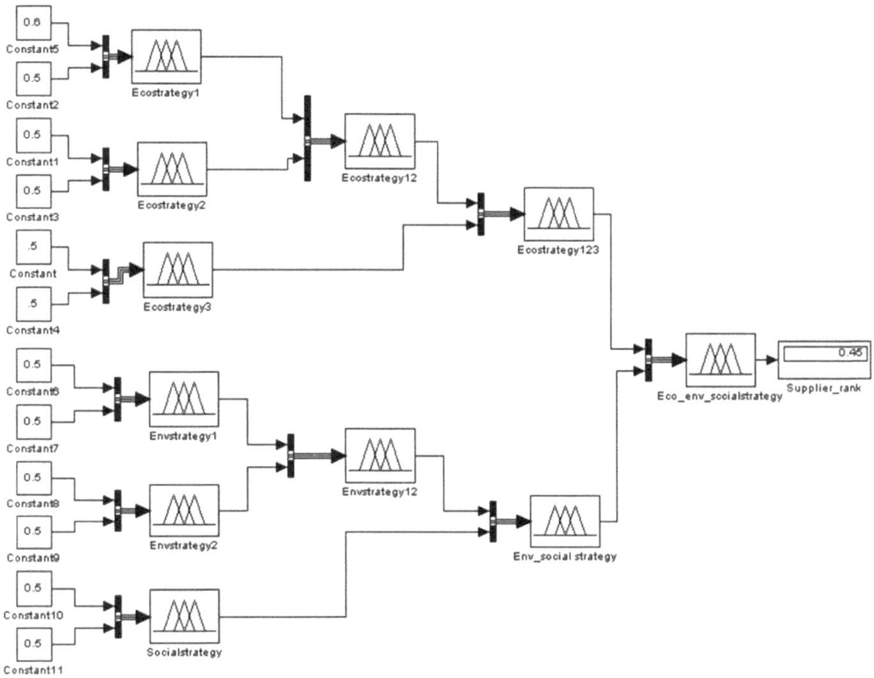

Fig. 4.15 Simulink model for sustainable supplier selection

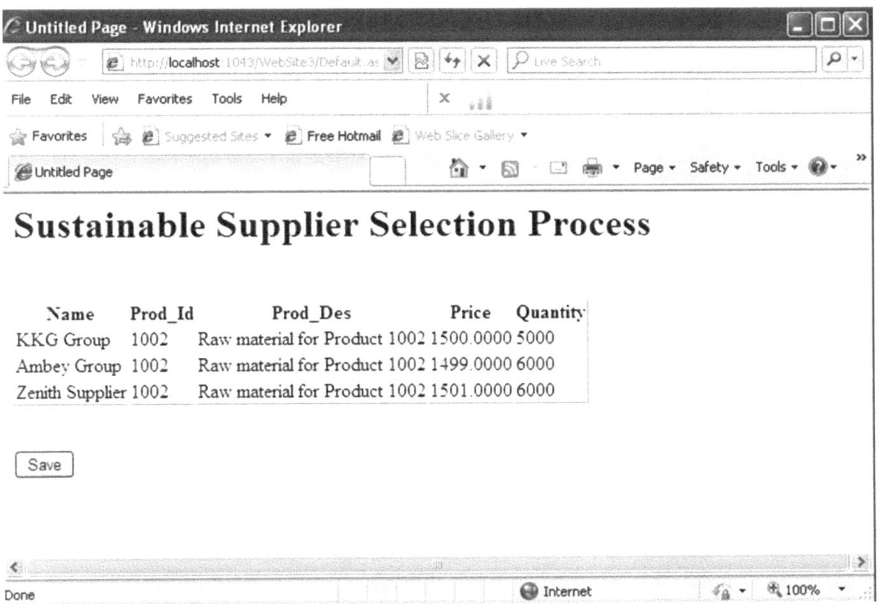

Fig. 4.16 Web-based sustainable supplier selection process

Table 4.6 Volume discount schedule

r	Volume discount (in thousand Rs)	Percent discount
1	0 to under 10	0
2	10 to under 20	10
3	20 and above	20

Table 4.7 Supplier datasheet

Sl No.	% failure rate of supply (f)	Reliability ($\alpha = 1 - f$)	Capacity (Ton)	Total cost (Rs/Ton)
1	5	0.95	5000	1500
2	5	0.95	6000	1499
3	4	0.96	6000	1501

Cost constraint: $1500 * x1 + 1499 * x2 + 1501 * x3 \leq 30,000,000$

Above multi-objective mixed integer non-linear constrained optimization (MOMINLP) problem is solved with MATLAB GA solver. Minimum value of TCP is 1.4396e + 007 and maximum value of TVRP is 5.4341e + 003. Best value of decision variables are {4.9993e + 03, 5.8943e + 03,1.1064 + e03}. Obtained pareto front is shown in Fig. 4.17. The average distance measure of the solutions on

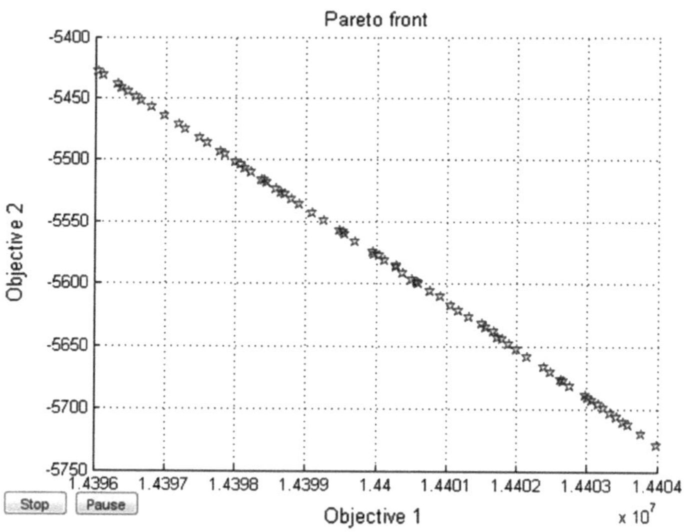

Fig. 4.17 Pareto front of TCP and TVRP

the pareto front was 0.0105462, and the spread measure of the pareto front was 0.161515.

4.7 Strategic Sourcing of Large Number of Suppliers: An Illustrative Case

Success begins with right decision. Right decision gives better insight, and better insight helps to frame effective strategy to mitigate future risk, if any. Strategic sourcing is a complex decision-making problem, and it demands better insight to control each parameter at microlevel of a process or system to prevent future risk. Strategic supplier selection encompasses operational metrics such as cost, quality, delivery, and other strategic dimensions such as quality management practices, process capabilities, design and development capabilities, and cost reduction capabilities into the decision-making process (Talluri and Narasimhan 2004). Strategic supplier selection demands more effective method to identify the difference in performance of suppliers across supplier groups to give better insights to decision makers. Efficiency of cascaded FIS solely depends on various criteria, namely structure of the membership functions and complexity of the rule base. Improper selection of membership function may lead to a disaster. Therefore, more flexible method is required for the strategic selection of large number of suppliers. Integrated approach of **data envelopment analysis (DEA)** and **non-parametric test** proposed by Talluri and Narasimhan (2004) to cluster supplier for strategic sourcing is limited to small number of suppliers as their proposed method is time consuming. In this section, data clustering technique is discussed to cluster suppliers for strategic sourcing. Proposed method is implemented with "R" and spreadsheet. **Cluster analysis (CA)** is a statistical technique which is used for the classification of similar objects into different groups by partitioning of a given data set into subsets (clusters) to identify common trait of each group or cluster. Clustering is a method of unsupervised learning and a well-known technique for statistical data analysis. Clustering algorithms can be classified into various categories such as **hierarchical methods, partitioning techniques, grid-based methods, density-based methods, and model-based methods**. One of the biggest challenges of clustering is identifying the **optimal number of clusters** and the selection of appropriate clustering method. Thirty indices are identified in this regard to select optimal value of cluster (Charrad et al. 2014). Out of thirty indices, five outperform others (Milligan and Cooper 1985). Majority rule was used further to select optimal number of clusters. The steps of this method are mentioned below as follows:

1. Define strategic goal of sustainable procurement.
2. Select criteria for sustainable procurement process. Identify cost criteria and benefit criteria. Benefit criterion is preferred to be larger-is-better except the

cost criterion (smaller-is-better). Convert cost criteria as the difference between highest cost and the cost offered by *i*th supplier to use it as benefit criteria.

3. Prepare master datasheet from each supplier datasheet with a suitable scale. Master datasheet is usually prepared with the responses of suppliers. In such case, it is advisable to conduct the test of hypothesis to confirm that questionnaire is simple and unambiguous and respondent can answer all without any difficulty (Talluri and Narasimhan 2004). Conduct data cleaning procedure and check redundant data. Conducts spend analysis to select major suppliers, if required.
4. Calculate total score of each supplier w.r.t each criteria. Calculate weighted total score of each supplier with the priority of each criterion. Rank suppliers accordingly in descending order to show cross-performance of each supplier under each criterion.
5. Select optimal number of clusters.
6. Select suitable clustering techniques with cluster validity method.
7. Cluster master data sheet.
8. Interpret each cluster to give better managerial interpretation.
9. Identify best suppliers. Implement appropriate strategy to selected group of suppliers to achieve strategic goal. Identify key suppliers and prune worst performer from supply pool. Tier selected suppliers, if required.
10. Select appropriate order allocation model for selected suppliers.

4.7.1 Case Study

To maintain confidentiality of business and secrecy of data, name of the company is changed and case study is further modified to extend the work of Talluri and Narasimhan (2004). Company ABC has its manufacturing base in south India and willing to initiate strategic sustainable procurement process for its new factory situated in north India. Company ABC has adopted the decentralized procurement process and highly depends on its local suppliers. Company ABC has operations plants, R&D facilities, and excellent distribution facilities. The critical objectives of the Company ABC in procurement and supply management include the following:

- Selection of key suppliers;
- Building long-term relationship with key suppliers;
- Initiating benchmarking program to encourage ineffective suppliers to improve their performance;
- Identification of worst suppliers to reject them from supply base;
- Initiating strategic sustainable procurement process to fulfill company's long-term goal; and
- Initiating strategies to reduce supply base and willing to apply supply base reduction method.

Company ABC is also interested to implement an effective user friendly method to select suitable suppliers to fulfill its goal and to achieve it. Company ABC is willing to start bimonthly meeting with its suppliers to prepare them as per the requirement of the company. Purchasing department of the company in south India is willing to adopt supply base reduction method to reduce supply base to increase profit margin and its new factory in north India is giving more emphasize on the quality of their raw material for the new product. Therefore, uncertainty of demand is quite expected. Moreover, traditional spend analysis cannot be used as local suppliers are going to supply for first time. Overall reduction in supply base can be achieved initially by rationalizing supply base in each location of the Company ABC. In the next step, spend analysis as well as integration of common suppliers for both factories could be done. Such gradual approach of supply base reduction will help new factories of Company ABC to work effectively with the tempo of local suppliers. Sourcing is a product-specific process. Commodity strategy is a priori to supplier base reduction approach. Commodity strategy gives guideline for optimal number of suppliers, type of suppliers or desired suppliers, desired supplier performance for each component bought. Thus, Company ABC should develop commodity strategy for its each product, parts, or components. In the absence of commodity strategy, Company ABC can use **Kraljic matrix** which is a 2 × 2 matrix to compare profit impact with supply risk. However, Company ABC is willing to start supply base reduction immediately. Company is also unaware whether supply base reduction approach will bring any benefit to its stakeholders or not. It is true that reduction in supply base will help Company ABC to develop long-term relationship with its suppliers and to have more priority from its suppliers and thereby company can expect to have more purchase power. But uncertainties associated with the new product demands considerable number of suppliers to assure availability of raw material for production. Instead of minimum number of suppliers, company may look for optimal number of suppliers to have a better result. In order to achieve these goals, company has selected three metrics and twelve criteria. Instead of sending questionnaire to all suppliers, suppliers were asked to send detailed supplier datasheet as per the prescribed format of the company. Company ABC initially screened about 100 suppliers. For the sake of calculations, only 11 suppliers are considered here. A master datasheet was prepared with 0-1 scale as shown in Table 4.9. Prior to that, data redundancy check was also conducted for data cleaning purpose. To select the optimal number of clusters, different indices were used as shown in Table 4.8, Figs. 4.18 and 4.19. Initially, hierarchical method and partitioning techniques were used with same data, as shown in Table 4.9, and the best technique was identified as K-means clustering technique for proposed supplier selection problem by comparing the results.

Table 4.8 Selection of optimal number of clusters

	Index	Number of clusters
1	Duda	3
2	Calinski and Harabasz	4
3	d index	3
4	Hubert index	3

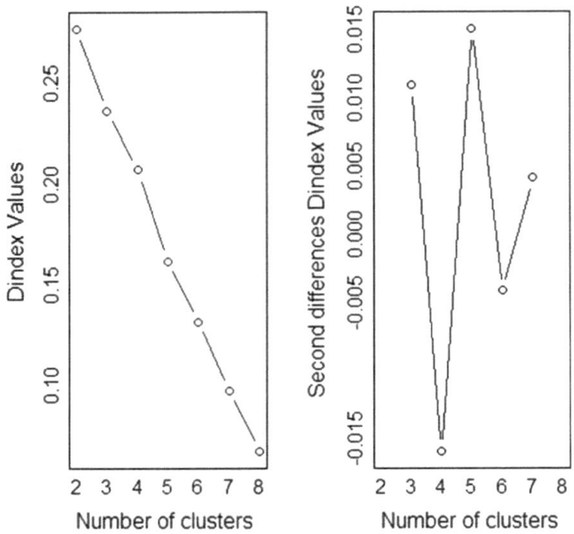

Fig. 4.18 d index

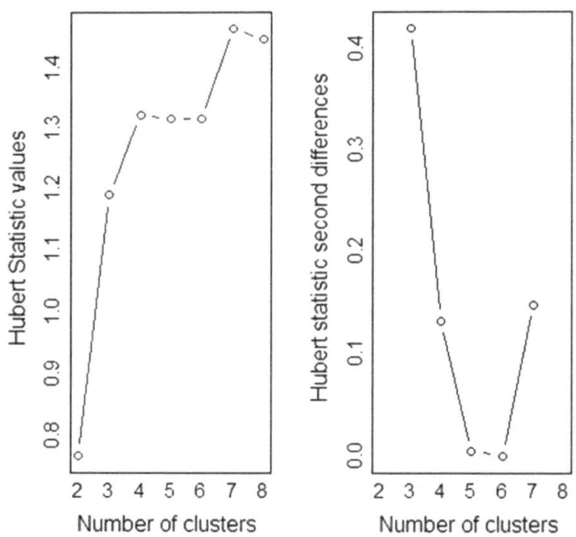

Fig. 4.19 Hubert index

K-means clustering technique categorized 11 suppliers into three different groups, namely best (B), medium (M), and worst (W) which is to be pruned as shown in Fig. 4.20.

Only members of the cluster 3 should be selected for order allocation stage. Performance of cluster 3 members should be benchmarked to encourage

Table 4.9 Master datasheet

			Economic strategy						Environmental strategy						Social				
Supp#	C1	C2	C3	C4	C5	C6	TS	Rank	C7	C8	C9	C10	TS	Rank	C11	C12	TS	Rank	WS
1	0.45	0.5	0.6	0.6	0.5	0.5	3.15	2	0.5	0.55	0.5	0.55	2.1	2.5	0.7	0.55	1.3	1	2.145
2	0.5	0.5	0.5	0.5	0.6	0.5	3.1	3	0.65	0.5	0.5	0.6	2.25	1	0.6	0.5	1.1	5	2.1285
3	0.7	0.5	0.35	0.5	0.5	0.5	3.05	4.5	0.45	0.5	0.45	0.5	1.9	6.5	0.7	0.5	1.2	1.5	2.0295
4	0.1	0.7	0.5	0.65	0.6	0.5	3.05	4.5	0.5	0.5	0.5	0.5	2	4	0.65	0.5	1.2	4	2.046
5	0.35	0.5	0.5	0.45	0.25	0.5	2.55	11	0.55	0.55	0.55	0.45	2.1	2.5	0.55	0.65	1.2	1.5	1.9305
6	0.5	0.6	0.45	0.55	0.6	0.5	3.2	1	0.45	0.45	0.5	0.5	1.9	6.5	0.55	0.45	1	7.5	2.013
7	0.65	0.35	0.5	0.35	0.35	0.5	2.7	9.5	0.5	0.45	0.5	0.5	1.95	5	0.5	0.5	1	7.5	1.8645
8	0.6	0.35	0.55	0.5	0.5	0.5	3	6	0.4	0.45	0.5	0.5	1.85	8	0.45	0.5	1	9	1.914
9	0.3	0.5	0.45	0.5	0.5	0.5	2.75	8	0.25	0.3	0.5	0.35	1.4	11	0.45	0.6	1.1	6	1.716
10	0.56	0.45	0.5	0.25	0.5	0.5	2.76	7	0.35	0.35	0.35	0.45	1.5	10	0.3	0.5	0.8	11	1.6698
11	0.2	0.65	0.35	0.5	0.5	0.5	2.7	9.5	0.5	0.5	0.3	0.5	1.8	9	0.4	0.45	0.9	10	1.7655

Note C1: Price, C2: Quality, C3: Service, C4: Delivery, C5: Capacity, C6: Past Performance C7: Environmental competencies, C8: Environmental management system, C9: Green packaging ability, C10: Green product design, C11: Health and safety C12: Education, TS: Total score, WS: Weighted Score

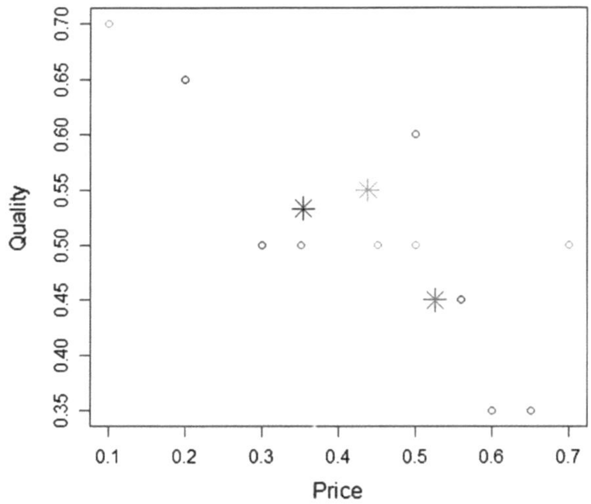

Fig. 4.20 Clusters with their centers

Table 4.10 Significance of three clusters

Cluster	Supplier	Remarks
3 (best)	1, 2, 3, 4	Cluster 3 groups all suppliers who have any combination of the following traits—low price, better quality, better delivery, better service, better capacity, good environmental competencies, better environmental management system, and well-documented policies and plans for safety, health and education for employees
2 (medium)	5, 6, 7, 8	Cluster 2 groups all suppliers who performed moderately w.r.t 12 sub-criteria and need more improvement
1 (worst)	9, 10, 11	All members of cluster 1 should be rejected

improvement of supplier 5, 6, 7, and 8. As shown in Tables 4.9 and 4.10, supplier 6 has better capacity and also offered better quality and prompt delivery. But he failed to get selected as he paid less attention to environmental and social metrics. Still suppliers 5 and 6 could be retained in supplier pool for future requirement as they performed moderately well. Long-term relationship should be maintained with all members of cluster 3, and special attention should be paid to suppliers 1 and 2 as they are the best performers among 11 suppliers. Suppliers 9, 10, and 11 should be pruned from supply pool. Such differentiation among suppliers performance gives better insight which is required for strategic sourcing. Tiering of suppliers are not considered as Company ABC is willing to pay equal priorities to all suppliers of cluster 3 and want to tier its supplier in near future. Finally, any appropriate allocation model stated earlier can be considered to allocate order to all members of cluster 3 as per the requirement of the Company ABC.

4.8 Conclusion

In this chapter, an attempt has been made to discuss strategic sustainable supplier selection process with three different methods—intuitionistic fuzzy AHP (IF-AHP), cascaded fuzzy inference system, and K-means data clustering method. Last two methods are discussed with an example and a case study. Strategic sourcing gives deeper insight to companies to take better decision to survive in cut throat competition. Intuitionistic fuzzy set is a generalized fuzzy set. It considers the degree of hesitation of decision makers. It is expected that proposed method could bring more flexibility to decision-making process for complex strategic sourcing problem. Proposed cascaded fuzzy inference system demands more resources such as MATLAB and .NET server. Cascaded FIS brings more computational complexity and depends highly on rule base. At the same time, structure of membership function is also subjective. Thus, it demands modification of rule base as well as modification of membership functions depends upon the need of the problem. The last method is more or less suitable for large number of suppliers and it gives better insight for strategic sourcing. This method can easily be implemented with Microsoft Excel. Companies do prefer simple, easy-to-understand method which they could implement easily with available resources such as Microsoft Excel. This is one of the reasons for acceptance of Saaty's AHP in all major decision-making processes. In this regard, author strongly suggests interested readers to refer work of Saaty on dynamic decision-making process. Our decision is not static. Our decision changes with time and place, and is usually influenced in the presence of external stimuli. However, judicious decision making is an art as selection of suitable method can give a better managerial interpretation of the problem and thereby an extra edge to the company to enhance its supply chain surplus.

References

Amindoust A, Ahmed S, Saghafinia A, Bahreininejad A (2012) Sustainable supplier selection: A ranking model based on fuzzy inference system. Appl Soft Comput 12:1668–1677

Bai C, Sarkis J (2010) Green supplier development: Analytical evaluation using rough set theory. J Clean Prod. doi:10.1016/j.jclepro.2010.01.016

Brink S, Diehl JC, Stevels A (1998) ECO-QUEST, an ecodesign self audit tool for suppliers of the electronics industry. In: Proc. the 1998 international symposium on electronics and the environment, May 4–6, Oak Brook, Illinois, pp 129–132

Carr AS, Pearson JN (1999) Strategically managed buyer–seller relationships and performance outcomes. J Oper Manag 17:497–519

Charrad M, Ghazzali N, Boiteau V, Niknafs A (2014) NbClust: An R package for determining the relevant number of clusters in a data set. J Stat Softw 61(6):1–36

Chen TY (2011) A comparative analysis of score functions for multiple criteria decision making in intuitionistic fuzzy settings. Inf Sci 181:3652–3676

Council for solid waste solution (1990). www.americanchemistry.com/s_plastics/bin.asp?CID=1211&DID=4601&DOC=FILE.PDF

Das SK, Mathew S (1999) Characterization of material outputs from an electronics demanufacturing facility. In: proc. IEEE international symposium on electronics and the environment, Boston, MA, May 11–13, pp 251–256

Das SK, Naik S (2002) Process planning for product disassembly. Int J Prod Res 40(6):1335–1355

Das S, Yedlarajiah D (2002) An integer programming model for prescribing material recovery strategies. In: Proc. IEEE international symposium on electronics and the environment, May 6–9, pp 118–122

De Toni A, Nassimbeni G (1999) Buyer–supplier operational practices, sourcing policies and plant performance: result of an empirical research. Int J Prod Res 37(3):597–619

Franke J (1995) Political evolution of EMAS: perspectives from the EU, National governments and industrial groups. Bus Strategy Environ 5(3):14–17

Franklin Associates (1991) Product life-cycle assessment: guidelines and principles. EPA Report, #68-CO-0003

Ghodsypour SH, O'Brien C (1998) A decision support system for supplier selection using an integrated analytic hierarchy process and linear programming. Int J Prod Econ 56–57:199–212

Guan Z, Jin Z, Zou B (2007) A Multi-Objective Mixed-Integer Stochastic Programming Model for the Vendor Selection Problem under Multi-Product Purchases. Inform Management Sci 18 (3):241–252

Gungor A, Gupta SM (1999) Issues in environmentally conscious manufacturing and product recovery: A survey. Comput Ind Eng 36:811–853

Gupta M (1995) Environmental management and its impact on the operations function. Int J Oper Prod Manage 15(8):34–51

Handfield R, Walton S, Sroufe R, Melnyk S (2002) Applying environmental criteria to supplier assessment: a study in the application of the analytical hierarchy process. Eur J Oper Res 141:70–87

Hsu C-W, Hu AH (2009) Applying hazardous substance management to supplier selection using analytic network process. J Clean Prod 17:255–264

Hunt R, Sellers J, Franklin W (1992) Resource and environmental profile analysis: a life cycle environmental assessment for products and procedures. Environ Impact Assess Rev 12(3): 245–269

Ishii K, Eubanks CF, Marco PD (1994) Design for product retirement and material life-cycle. Mater Des 15(4):225–233

Kim K, Song I, Kim J, Jeong B (2006) Supply planning model for remanufacturing system in reverse logistics environment. Comput Ind Eng 51:279–287

Kotha S (1995) Mass customisation: implementing the emerging paradigm for competitive advantage. Strategic Manage J 16:21–42

Lee HI, Kang HY, Hsu CF, Hung HC (2009) A green supplier selection model for high-tech industry. Expert Syst Appl 36(4):7917–7927

Levan LS (1998) Life cycle assessment: measuring environmental impact. www.fpl.fs.fed.us/documnts/pdf1998/levan98b.pdf

Milligan GW, Cooper MC (1985) An examination of procedures for determining the number of clusters in a data set. Psychometrika 50(2):159–179

Mukherjee K, Sarkar B, Bhattacharya A (2011) Comments on the erratum to "Supply planning model for remanufacturing system in reverse logistics environment" [Comput. Ind. Eng. 51 (2006) 279–287]. Comput Ind Eng 61:1349–1350

Muralidharan C, Anantharaman N, Deshmukh (2002) A multi-criteria group decision making model for supplier rating. J Supply Chain Manage Fall 22–33

Noci G (1997) Design "green" vendor rating systems for the assessment of a supplier's environmental performance. Eur J Purchasing Supply Manage 3(2):103–114

Pine BJ (1993) Mass customisation. HBS Press, Boston

Sanchez PP, Soyer R (1998) Information concepts and pair-wise comparison matrices. Inf Processing Lett 68:185–188

Shin H, Collier DA, Wilson DD (2000) Supply management orientation and supplier buyer performance. J Oper Manage 18:317–333

Smith KG, Carroll SJ, Ashford SJ (1995) Intra-and inter-organizational cooperation: toward a research agenda. Acad Manag J 38(1):7–23

Spekman RE (1988) Perceptions of strategic vulnerability among industrial buyers and its effect on information search and supplier evaluation. J Bus Res 17:313–326

Stilwell JR, Canty PK, Montrone A (1991) Packaging for the environment. American Management Association, New York

Svoboda S (1995) Note on life cycle analysis www.umich.edu/~nppcpub/resources/compendia/CORPpdfs/CORPlca.pdf

Szmidt E, Kacprzyk J (2000) Distance between intuitionistic fuzzy sets. Fuzzy Sets, Syst 114 (3):505–518

Tracey M, Tan CL (2001) Empirical analysis of supplier selection and involvement, customer satisfaction and firm performance. Supply Chain Manage: Int J 6(4):174–188

Van Hoek RI (1999) From reversed logistics to green supply chains. Supply Chain Manage: Int J 4 (3):129–135

Wang M, Perkins JR (2006) Using interval alignment policies for efficient production control of supply chain systems. Int J Ind Syst Eng 1(1–2):87–108

Wang M, Perkins JR (2011) Time interval alignment (ia) policies, boundary and applications with multiple stream arrivals. J Syst Sci Syst Eng 20(4):400–415

Wu CH, Kuo TC, Lu YY (2007) Environmental principles applicable to green supplier evaluation by using multi-objective decision analysis. Int J Prod Res 45(18–19):4317–4331

Xia W, Wu Z (2007) Supplier selection with multiple criteria in volume discount environments. Omega 35:494–504

Zeid I, Gupta SM (2002) Computational algorithm to evaluate product disassembly cost index. In: Proc. of the SPIE international conference on environmentally conscious manufacturing II, February 11, pp 23–31

Zhang HC, Li J, Shrivastava P, Whitely A, Eugene M (2004) A web-based system for reverse manufacturing and product environmental impact assessment considering end-of-life dispositions. CIRP Annals-Manuf Technol 53(1):5–8

Chapter 5
A Note on Limitations of FAHP

> One of the most difficult tasks in multiple criteria decision analysis (MCDA) is determining the weights of individual criteria so that all alternatives can be compared based on the aggregate performance of all criteria.
>
> Chiang Kao
>
> Applied Mathematical Modelling (2010), Vol. 34, pp. 1779–1787

5.1 Introduction

In decision-making process, uncertainty can be classified into two categories, namely internal and external. External uncertainties refer to imprecise data, vague problem definition, etc. Internal uncertainties refer to subjectivity of the decision maker. Fuzzy set theory is a well-established subject, and its application is well accepted in engineering. But application of fuzzy set theory in decision sciences is not unanimously accepted yet. Fuzziness may work better in metric spaces than in order spaces (Saaty and Tran 2007). Particularly, fuzzy AHP has certain limitations (Saaty and Tran 2007, 2010; Zhu 2014, 2012). In 1986, Saaty, the originator of AHP, proposed four axiomatic foundations of AHP. Some of them are as follows:

1. **Reciprocal Axiom**: For all judgments a_{ij}, its reciprocal a_{ji}, must satisfy the following relation

$$a_{ij} \cdot a_{ji} = 1 \quad \forall i, j = 1,2,\ldots,n \tag{5.1}$$

$$\text{or } a_{ij} = \frac{1}{a_{ji}}$$

Fuzzy AHP basically violates the above relation. For example, a triangular fuzzy number $\widetilde{a_{ij}} = (2,3,4)$ and its reciprocal $\widetilde{a_{ji}} = (\frac{1}{4}, \frac{1}{3}, \frac{1}{2})$

K. Mukherjee, *Supplier Selection*, Studies in Systems, Decision and Control 88,
DOI 10.1007/978-81-322-3700-6_5

$$\widetilde{a_{ij}} \otimes \widetilde{a_{ji}} = (l, m, u) \otimes (1/u,\ 1/m,\ 1/l) \simeq (l/u,\ m/m,\ u/l) \neq (1, 1, 1) \quad (5.2)$$

where $\otimes$ represents fuzzy multiplication.
The above multiplication approach was used by Chang (1996). Van Laarhoven and Pedrycz (1983) used following approach:

$$(l1,\ m1,\ u1) \otimes (l2,\ m2,\ u2) \sim (l1\ l2,\ m1\ m2,\ u1\ u2) \neq (1, 1, 1) \quad (5.3)$$

The above multiplication is an approximation.

2. **Consistency Axiom**: In AHP, if all judgments are consistent, then following relation must satisfy

$$a_{ik} \cdot a_{kj} = \mathrm{a_{ij}} \quad \forall i, j, k = 1, 2, 3, \ldots., n \quad (5.4)$$

A positive fuzzy reciprocal matrix $\tilde{A} = [\widetilde{a_{ij}}]$ is consistent if and only if $\widetilde{a_{ik}} \otimes \widetilde{a_{kj}} = \widetilde{a_{ij}}$.
For example, if $\widetilde{a_{11}} = (1, 1, 1)$; $\widetilde{a_{12}} = (2, 3, 4)$ and $\widetilde{a_{21}} = (1/4,\ 1/3,\ 1/2)$, then

$$\widetilde{a_{12}} \otimes \widetilde{a_{21}} \neq \widetilde{a_{11}}$$

3. **Continuity Axiom**: If $A \succ B \succ C$, then there exist probability p in (0, 1) such that $B \sim pA + (1 - p)C$ (Zhu 2014). For example, $M1$(4, 5, 6), $M2$(2, 3, 4), $M3$ (1, 1, 1)

$$M1 \succ M2 \succ M3.$$

Therefore,

$$M2 \sim pM1 + (1 - p)M3 \quad (5.5)$$

The above equation can be written as follows:

$$\begin{cases} 2 \sim 4p + (1 - p) \\ 3 \sim 5p + (1 - p) \\ 4 \sim 6p + (1 - p) \end{cases} \quad (5.6)$$

$$\begin{cases} p \sim 1/3 \\ p \sim 1/2 \\ p \sim 3/5 \end{cases} \quad (5.7)$$

None of the above value of p satisfies Eq. (5.6) Thus, no p in (0, 1) satisfies Eq. (5.7). Hence, fuzzy AHP violates axiom of reciprocal, axiom of consistency, and axiom of continuity.
Zhu (2014) mentioned that Saaty's fundamental 1–9 scale itself is fuzzy. Terms used in fundamental scale such as 'moderate importance' and 'strong

importance' are linguistic variables to express degree of dominance of one variable to other. Thus, values of fundamental scale are not crisp, and there is absolute no need of fuzzification of fundamental scale for fuzzy AHP. Saaty and Tran (2007) mentioned that good judgment produces valid answer in AHP, and fuzzy AHP simply produces perturbation without producing any better outcomes. Even inconsistent good judgment yields better result in AHP, but fuzzy AHP produces worse rather than better outcomes.

5.2 Other Limitations of Fuzzy AHP Models

Some of the limitations of the Van Laarhoven and Pedrycz (1983) method are as follows:

- Methodology proposed by them to normalize the local fuzzy weights was problematic (Wang and Parkan 2006).
- Their proposed method for incomplete fuzzy comparison matrices for local fuzzy weights was not suitable (Wang and Parkan 2006).

Some of the limitations of Buckley (1985) method are as follows:

- If the pairwise comparison matrix is consistent, then their method yields same priority vectors as the eigenvectors method, which was Saaty's classical AHP method. But it fails to produce same result for inconsistent matrix (Csutora and Buckley 2001).

Some the limitations of Chang (1996) model are as follows:

- Proposed method of Chang could lead to a wrong decision if it assigns zero weights to some items such as criteria, sub-criteria, or alternatives and excludes them from the decision analysis (Wang et al. 2008).

5.3 Consistency Index and Optimization Methods for AHP

Saaty proposed consistency ratio (C.R) which is the ratio of consistency index (C.I) and random index (R.I) and defined as follows:

$$\text{C.I} = \frac{\lambda_{\max} - n}{n - 1} \quad \text{and} \quad \text{C.R} = \frac{\text{C.I}}{\text{R.I}} < 0.1$$

Thus C.R can be simplified as follows:

$$\mathrm{C.R} = \frac{\lambda_{\max} - n}{R(n-1)} < 0.1 \Rightarrow \lambda_{\max} < 0.1.R\,(n-1) + n \tag{5.8}$$

With the help of consistency index Saaty imposed an extra constraint on principal eigenvalue, shown in Eq. 5.8, to reduce values of all elements of pairwise comparison matrix except its diagonal elements. For example, values of a_{12}, a_{13}, and a_{23} should be reduced through iteration so that a 3×3 pairwise comparison matrix could satisfy consistency index or maximum limit of principal eigenvalue, $\lambda_{\max}$, shown in Table 5.1.

There are several other methods are available to derive priorities from AHP. In this sec., optimization approach is considered mainly.

Optimization Approaches to find priority for AHP.

5.3.1 Weighted Least Square Method (Chu et al. 1979)

$$a_{ij} \approx \frac{W_i}{W_j} \tag{5.9}$$

$$\text{Min } S = \sum_{i=1}^{n} \sum_{j=1}^{n} (a_{ij} - W_i / W_j)^2 \tag{5.10}$$

Subject to

$$\sum_{i=1}^{n} W_i = 1 \tag{5.11}$$

5.3.2 Error Minimization Method (Chen and Triantaphyllou 2001)

$$a_{ij} = \frac{W_i}{W_j}(1 + \varepsilon_{ij}) \tag{5.12}$$

Equation 5.10 should be modified with Eq. 5.11.

Table 5.1 Limit of principal eigenvalue (RI is from Saaty and Tran 2007; with kind permission from Elsevier Limited)

Size, n	1	2	3	4	5	6	7	8	9	10	11	12	13	14
Random index	0	0	0.52	0.89	1.12	1.26	1.36	1.41	1.46	1.49	1.52	1.54	1.6	1.58
Principal eigenvalue is less than equal to	1	2	3.104	4.267	5.448	6.63	7.816	8.987	10.17	11.34	12.52	13.69	14.9	16.05

5.3.3 Logarithmic Least Square Method

$$\text{Min} \sum_{i=1}^{n} \sum_{j>i}^{n} (\ln a_{ij} - (\ln W_i - \ln W_j))^2 \tag{5.13}$$

Subject to

$$\sum_{i=1}^{n} W_i = 1 \tag{5.14}$$

5.3.4 Goal Programming Method (Bryson 1995)

$$a_{ij} = \frac{W_i}{W_j}\left(\frac{\delta_{ij}^{+}}{\delta_{\overline{ij}}}\right) \quad \forall i, j = 1, 2, 3, 4, \ldots, n \text{ and } j > i \tag{5.15}$$

where $\delta_{ij}^{+} \geq 1$ and $\delta_{ij}^{-} \geq 1$ are deviation variables and both of them cannot be greater than 1 simultaneously.

$$\text{Min} \sum_{i=1}^{n} \sum_{j>i}^{n} (\ln \delta_{ij}^{+} + \ln \delta_{ij}^{-}) \tag{5.16}$$

Subject to

$$\ln a_{ij} = inW_i - inW_j + \left(\ln \delta_{ij}^{+} - \ln \delta_{ij}^{-}\right) \tag{5.17}$$

Unfortunately, none of the optimization method considered maximum limit of principal eigenvalue, shown in Table 5.1.

5.4 Alternative Approaches to FAHP

Saaty mentioned that fuzzy AHP brings more fuzziness in the result. To explain Saaty's statement, let us consider two triangular fuzzy numbers (TFNs) $\widetilde{A} = \{1, 2, 3\}$ and $\widetilde{B} = \{2, 3, 4\}$ are used for pairwise comparison of criteria or alternatives. TFNs are considered in this regard, as they are commonly used in fuzzy AHP. The product of the above two TFNs are equal to {2, 6, 12}. As mentioned in Chap. 1, triangular fuzzy number $\widetilde{A}$ refers to the linguistic term 'about to 2'. If we decrease differences between lower and upper limit of TFN, then we will become closer to average value. It means certainty increases with the decrease in support of any TFN or alternatively, we can say multiplication of TFNs brings more fuzziness

Table 5.2 Method of Triantaphyllou and Lin (1996)

	C_1	C_2	C_3	Calculation of priority
C_1	(1, 1, 1)	(5/2, 3, 7/2)	(7/2, 4, 9/2)	(35/4, 12, 63/4) = $(8.75, 12, 15.75)^{1/3}$ = (2.0606, 2.2894, 2.5066) = (0.187, 0.208, 0.2278)
C_2	(2/7, 1/3, 2/5)	(1, 1, 1)	(3/2, 2, 5/2)	(3/7, 2/3, 1) = $(0.428, 0.66,1)^{1/3}$ = (0.7536, 0.8707, 1.0000) = (0.0685, 0.079, 0.09)
C_3	(2/9, 1/4, 2/7)	(2/5, 1/2, 2/3)	(1, 1, 1)	(4/45, 1/8, 4/21) = $(0.088, 0.125, 0.19)^{1/3}$ = (0.4448, 0.5000, 0.5749) = (0.04, 0.045, 0.052)

to the result as multiplication increases support of the product which is also a TFN. Some of the methods for deriving priority for FAHP are as follows:

5.4.1 Method of Triantaphyllou and Lin (1996)

Triantaphyllou and Lin (1996) proposed an alternative approach to quantify priority of alternatives. Their proposed method is explained with an example, previously solved in Chap. 1 with Extent Fuzzy AHP (Table 5.2).

Obtained priorities show that $C_1 \succ C_2 \succ C_3$.

Both Extent Fuzzy AHP (Chang 1995) and method of Triantaphyllou and Lin (1996) did not consider consistency of fuzzy pairwise comparison matrix and their result significantly deviates from each other.

5.4.2 Least Square Distance Method (Wang and Parkan 2006; with Kind Permission from Elsevier Limited)

If $\widetilde{R}_i$ and $\widetilde{R}_j$ be the two triangular or trapezoidal numbers. The weighted Euclidean distance between $\widetilde{R}_i$ and $\widetilde{R}_j$ is defined as follows

$$d_{ij} = \sum_{k=1}^{m} \sqrt{(W_i r_{ik} - W_j r_{jk})^2} \tag{5.18}$$

$$\text{Min } S = \sum_{i=1}^{n} \sum_{j=1, j \neq i}^{n} d_{ij}^2 \tag{5.19}$$

Subject to

$$\sum_{i=1}^{n} W_i = 1 \quad (5.20)$$

5.4.3 Defuzzification-Based Least Square Method (Wang and Parkan 2006; with Kind Permission from Elsevier Limited)

If $\widetilde{R_i}$ and $\widetilde{R_j}$ be the two triangular or trapezoidal numbers then their defuzzified value Z can be expressed as follows:

For triangular fuzzy number, $Z = \frac{1}{3}(r_1 + r_2 + r_3)$.

For trapezoidal fuzzy number, $Z = \frac{1}{4}(r_1 + r_2 + r_3 + r_4)$.

The defuzzified values of the weighted fuzzy numbers should be as close as possible. Their proposed objective function is as follows:

$$\text{Min } S = \sum_{i=1}^{n} \sum_{j=1, j \neq i}^{n} (W_i z_i - W_j z_j)^2 \quad (5.21)$$

Subject to

$$\sum_{i=1}^{n} W_i = 1 \quad (5.22)$$

5.4.4 Preference Programming (Salo and Hämäläinen 1995)

Salo and Hämäläinen (1995) extended the work of Arbel (1993) and used series of optimization to derive priority vectors for criteria or alternatives from the feasible region of decision space. They divided entire hierarchy into several levels and used top-down or bottom-up or any feasible combination to derive priority. One of the major significance of their method is the graphical presentation of feasible region of decision space and introduction of ambiguity index (AI). Their method is included in this section as any triangular or trapezoidal fuzzy number can be easily converted to interval values with fuzzy alpha-cut method. But such generalization of fuzzy number is always associated with loss of information. Let us consider that a decision maker has prepared following pairwise comparison matrix for three alternatives w.r.t a given criterion

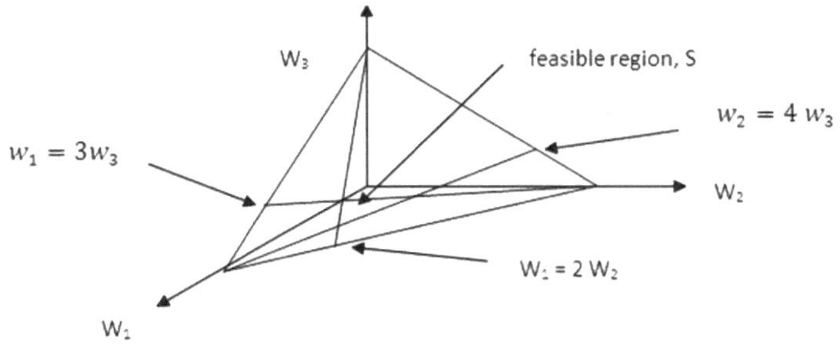

Fig. 5.1 Feasible region of decision space

$$S = \begin{bmatrix} 1 & 2 & 3 \\ 1/2 & 1 & 4 \\ 1/3 & 1/4 & 1 \end{bmatrix} = \begin{bmatrix} \frac{W_1}{W_1} & \frac{W_1}{W_2} & \frac{W_1}{W_3} \\ \frac{W_2}{W_1} & \frac{W_2}{W_2} & \frac{W_2}{W_3} \\ \frac{W_3}{W_1} & \frac{W_3}{W_2} & \frac{W_3}{W_3} \end{bmatrix} \tag{5.23}$$

By comparing above two matrices, we can derive the following relations:

$$w_1 = 2w_2; \quad w_1 = 3w_3; \quad w_2 = 4\,w_3 \text{ and normalized priority } V = (\frac{6}{11}, \frac{3}{11}, \frac{2}{11}) \tag{5.24}$$

Feasible region, S of the above problem is shown in Fig. 5.1. Salo's method can convert the above pairwise comparison matrix is as follows:

$$S = \begin{bmatrix} 1 & 2 & [3, 8] \\ 1/2 & 1 & [3/2, 4] \\ [\frac{1}{8}, 1/3] & [\frac{1}{4}, 2/3] & 1 \end{bmatrix} \quad \text{and} \quad \text{AI} = 0.1129 \tag{5.25}$$

Ambiguity Index of Salo and Hämäläinen (1995) determines fuzziness associated in assigning interval limit in fuzzy pairwise comparison matrix. But they did not consider any consistency index in their calculation. As a result, an interval matrix with lower value of ambiguity can have higher inconsistency. However, their proposed method is flexible enough to solve complex multi-level hierarchical problem.

5.4.5 Fuzzy Preference Programming (Mikhailov and Singh 2003)

Fuzzy Preference Programming (FPP) converts prioritization problems into a fuzzy linear programming problem to identify non-convex, non-empty, fuzzy feasible

Table 5.3 Comparison of AHP and different FAHP methods

	C_1	C_2	C_3	Saaty's AHP	Triantaphyllou and Lin	Extent Fuzzy AHP	FPP
C_1	1	3	4	0.625	0.2076	1	0.62203
C_2	1/3	1	2	0.238	0.0792	0	0.23511
C_3	1/4	1/2	1	0.137	0.0457	0	0.14286

region that contains all priorities which satisfies all interval constraints (Mikhailov and Singh 2003). In FPP, decision makers consider a linear membership function to express degree of violation of the constraint mentioned below:

$$l_{ij} \tilde{\leq} w_i / w_j \tilde{\leq} u_{ij} \tag{5.26}$$

where $\tilde{\leq}$ denotes fuzzy less or equal.

The optimization problem prepared with the above constraint can be expressed as follows:

Maximize λ
subject to

$$(m_{ij} - l_{ij})\,\lambda w_j - w_i + l_{ij}w_j \leq 0 \quad \forall i = 1, 2, 3, \ldots, n-1;\ j = 2, 3, 4, \ldots, n \text{ and } j > i \tag{5.27}$$

$$(u_{ij} - m_{ij})\,\lambda w_j + w_i - u_{ij}w_j \leq 0 \tag{5.28}$$

$$\sum_{k=1}^{n} w_k \quad \forall k = 1, 2, 3, \ldots, n \tag{5.29}$$

For example, above problem, previously solved in Chap. 1 with Extent Fuzzy AHP, is solved with FPP (Table 5.3).

Result obtained from FPP, Extent Fuzzy AHP, and method of Triantaphyllou and Lin (1996) differs considerably and raises questions to select best method for the simple problem. Result of FPP is very close to the result of Saaty's AHP. It proves that FPP can give more accurate result as it does not consider multiplicative relation in its entire computation process. Today, more research work is required to well define consistency index of FAHP. In this regard, interested readers can refer work of Leung and Cao (2000).

5.5 Conclusion

Selection of appropriate MCDA tools is an open issue. Purpose of this chapter is to aware researchers about the limitations of some of the fuzzy AHP methods. It is crystal clear that all three models of fuzzy AHP, namely method of Van Laarhoven and Pedrycz (1983), Buckley (1985), and Chang (1996) violate axioms of Saaty's classical AHP. Among all methods, Fuzzy Preference Programming (Mikhailov and Singh 2003) gives result closer to the classical AHP. Thus, selection of an appropriate method is a multi-criteria problem to trade-off conflicting criteria such as accuracy of result, simplicity of method, and consistency of result.

References

Bryson N (1995) A goal programming method for generating priorities vectors. J Oper Res Soc 641–648

Buckley JJ (1985) Fuzzy hierarchical analysis. Fuzzy Sets Syst 17:233–247

Chang D-Y (1996) Applications of the extent analysis method on fuzzy AHP. Eur J Oper Res 95 (3):649

Chu ATW, Kalaba RE, Spingaran K (1979) A comparison of two methods for determining the weights of belonging to fuzzy sets. J Optim Theory Appl 27(4):321–538

Csutora R, Buckley JJ (2001) Fuzzy hierarchical analysis: the Lamda-Max method. Fuzzy Sets Syst 120:181–195

Mikhailov L, Singh MG (2003) Fuzzy Analytic Network Process and its Application to the Development of Decision Support Systems. IEEE Trans Syst, Man, Cybern Part C 33(1):33–41

Saaty TL, Tran LT (2007) On the invalidity of fuzzifying numerical judgments in the Analytic Hierarchy Process. Math Comput Model 46:962–975

Saaty TL, Tran LT (2010) Fuzzy Judgments and Fuzzy Sets. Int J Strateg Decis Sci 1(1):23–40

Salo A, Hämäläinen RP (1995) Preference programming through approximate ratio comparisons. Eur J Oper Res 82:458–475

Van Laarhoven PJM, Pedrycz W (1983) A fuzzy extension of Saaty's priority theory. Fuzzy Sets Syst 11:229–241

Wang Y-M, Parkan C (2006) Two new approaches for assessing the weights of fuzzy opinions in group decision analysis. Inf Sci 176:3538–3555

Wang Y-M, Luo Y, Hua Z (2008) On the extent analysis method for fuzzy AHP and its applications. Eur J Oper Res 186:735–747

Zhu K (2012) The invalidity of triangular fuzzy AHP—a mathematical justification. http://dx.doi.org/10.2139/ssrn.2011922

Zhu K (2014) Fuzzy analytic hierarchy process: fallacy of the popular methods. Eur J Oper Res 236(1):209–217

Appendix
MCDA Tools and Meta-Heuristic Techniques: Sample Codes

Computer Codes

A.1 MATLAB code for Analytic Hierarchy Process (AHP)

```
*********************************************************************
                    This code is developed on the algorithm proposed by T.L.Saaty for AHP

*********************************************************************
tic;
clc;
clear % is must required to clean workspace data %
row = input('Enter no of rows or columns :');
for i = 1:1:row
 for j= 1:1:row
  if i==j
   a(i,j)=1;
      elseif (i>j)
            a(i,j)= 1/a(j,i) ;
  else
    a(i,j)=input('Enter pair wise comparison value :');

  end;

 end
 end
disp('Pair Wise Comparison Matrix')
disp(a)

*************************************************************
% Normaliztion of pair wise matrix %

for i=1:1:row
  Sum_Col=0.0;
  for j=1:1:row
    Sum_Col= Sum_Col + a(j,i);
  end
    Col_Sum(i)= Sum_Col;
end
disp(Col_Sum)
for i=1:1:row
  for j=1:1:row
    norm_mat(j,i)= a(j,i)/Col_Sum(i);
  end
end
```


K. Mukherjee, *Supplier Selection*, Studies in Systems, Decision and Control 88,
DOI 10.1007/978-81-322-3700-6

```
disp('Normalized Matrix')
disp(norm_mat)
% Take average of row to calculate priority %
for i=1:1:row
   row_sum = 0.0;
      for j=1:1:row
            row_sum = row_sum + norm_mat(i,j);
      end
   prio_mat(i)= row_sum/row;
end
disp('Priority Vectors')
disp(prio_mat)
***********************************************
% Calculation of consistency index %

lambda_max=0.0;
for i=1:1:row
   lambda_max= lambda_max + Col_Sum(i)* prio_mat(i);
end
CI = (lambda_max - row) /(row -1);
disp('Consistency Index')
disp(CI)
% Random Consistency Index Formula%
RI=1.98*(row-2)/row;
disp('Random Consistency Index')
disp(RI);
CR=CI/RI;
disp('Consistency Ratio')
disp(CR)

***********************************************
%Pair Wise Comparison Matrix for Alternatives%

Alt_row =input('Enter No. of Alternatives:');
for k=1:1:row  %Comparison with respect to each criterion%
   disp(['Comparison with respect to Criteria :  ', num2str(k)])
   for i=1:1:Alt_row
   for j=1:1:Alt_row
  if i==j
   alt(i,j,k)=1;
       elseif (i>j)
              alt(i,j,k)= 1/alt(j,i,k) ; % Array of Matrix representation %
   else
      alt(i,j,k)=input('Enter pair wise comparison value :');

  end;
   end
   end
end
for i=1:1:row
   disp(alt(:,:,i))
end

*****************************************************
% Normaliztion of pair wise matrix %
```

```
for k=1:1:row
for i=1:1:Alt_row
   Sum_Col=0.0;
   for j=1:1:Alt_row
      Sum_Col= Sum_Col + alt(j,i,k);
   end
      Col_Sum_alt(k,i)= Sum_Col;
end
end

for i=1:1:row
disp(Col_Sum_alt(i,:))

end

for k=1:1:row
for i=1:1:Alt_row
   for j=1:1:Alt_row
      norm_mat_alt(j,i,k)= alt(j,i,k)/Col_Sum_alt(k,i);
   end
end
end
disp('Normalized Alternative Matrix')
for i=1:1:row
   disp(norm_mat_alt(:,:,i))
end

************************************************
 % Take average of row to calculate priority %

for k=1:1:row
for i=1:1:Alt_row
   row_sum = 0.0;
      for j=1:1:Alt_row
            row_sum = row_sum + norm_mat_alt(i,j,k);
      end
   prio_mat_alt(k,i)= row_sum/Alt_row;
end
end

disp('Local Priority Vectors Of Alternative')
for i=1:1:row
 disp(prio_mat_alt(i,:))
end

*************************************************
% Calculation of CR of each pair wise comparison matrix of alternatives %

for i=1:1:row
   lambda_max_alt=0.0;
   for j=1:1:Alt_row
   lambda_max_alt= lambda_max_alt + Col_Sum_alt(i,j)* prio_mat_alt(i,j);
   end
   lambda_max_new(i)= lambda_max_alt;
end
for i=1:1:row
CI_alt(i) = (lambda_max_new(i) - Alt_row) /(Alt_row -1);
end
```

```
disp('Consistency Index of Pairwise Comparion Matrix of Alternatives')
disp(CI_alt)

*****************************************
% Random Consistency Index Formula%

RI=1.98*(Alt_row-2)/Alt_row;
disp('Random Consistency Index')
disp(RI);
for i=1:1:row
CR_alt(i)=CI_alt(i)/RI;
end
disp('Consistency Ratio of Pairwise Comparison Matrix of Alternatives')

disp(CR_alt)

**********************************************
 %disp('Global Priority Vectors Of Alternative')

for i=1:1:Alt_row
   Prio_Sum =0.0;
  for j=1:1:row
   Prio_Sum = Prio_Sum + prio_mat(j)* prio_mat_alt(j,i);
  end
  global_prio(i)=Prio_Sum;
end
disp('Rank of Alternativies:')
sort_global_prio = sortrows(global_prio);
for i=1:1:Alt_row
      for j=1:1:Alt_row
if sort_global_prio(i)==global_prio(j)
   rank(j)= i;
end
   end
end
for i=1:1:Alt_row
disp(['Alternative : ', num2str(i),' Global Priority: ',num2str(global_prio(i)), '  Rank  :  ',num2str(rank(i))])
end
toc;
```

A.2 MATLAB Code for Fuzzy AHP by alpha-cut method

```
******************************************************************************
                                        This code is developed on fuzzy AHP alpha–cut method

******************************************************************************
tic;
clc;
clear % is must required to clean workspace data %
row = input('Enter no of rows or columns :');
for i = 1:1:row
 for j= 1:1:row
   if i==j
    a(i,j)=1;
        elseif (i>j)
              a(i,j)= 1/a(j,i) ;
```

```
        else
            a(i,j)= Fuzzy_Alpha_Cut();

        end;

    end
    end
disp('Pair Wise Comparison Matrix')
disp(a)

****************************************
 % Normaliztion of pair wise matrix %

for i=1:1:row
    Sum_Col=0.0;
    for j=1:1:row

        Sum_Col= Sum_Col + a(j,i);
    end
        Col_Sum(i)= Sum_Col;
end
disp(Col_Sum)
for i=1:1:row
    for j=1:1:row
        norm_mat(j,i)= a(j,i)/Col_Sum(i);
    end
end
disp('Normalized Matrix')
disp(norm_mat)

**********************************************
% Take average of row to calculate priority %

for i=1:1:row
    row_sum = 0.0;
        for j=1:1:row
                row_sum = row_sum + norm_mat(i,j);
        end
    prio_mat(i)= row_sum/row;
end
disp('Priority Vectors')
disp(prio_mat)

***********************************************
% Calculation of consistency index %

lambda_max=0.0;
for i=1:1:row
    lambda_max= lambda_max + Col_Sum(i)* prio_mat(i);
end
CI = (lambda_max - row) /(row -1);
disp('Consistency Index')
disp(CI)

**********************************************
% Random Consistency Index Formula%
RI=1.98*(row-2)/row;
disp('Random Consistency Index')
```

```
disp(RI);
CR=CI/RI;
disp('Consistency Ratio')
disp(CR)

****************************************************
%Pair Wise Comparison Matrix for Alternatives%

Alt_row =input('Enter No. of Alternatives:');
for k=1:1:row  %Comparison with respect to each criterion%
   disp(['Comparison with respect to Criteria :  ', num2str(k)])
   for i=1:1:Alt_row
   for j=1:1:Alt_row
 if i==j
   alt(i,j,k)=1;
       elseif (i>j)
               alt(i,j,k)= 1/alt(j,i,k) ; % Array of Matrix representation %

   else
      alt(i,j,k)= Fuzzy_Alpha_Cut();

 end;
   end
   end
end
for i=1:1:row
   disp(alt(:,:,i))
end

****************************************************
% Normaliztion of pair wise matrix %

for k=1:1:row
for i=1:1:Alt_row
   Sum_Col=0.0;
   for j=1:1:Alt_row
      Sum_Col= Sum_Col + alt(j,i,k);
   end
      Col_Sum_alt(k,i)= Sum_Col;
end
end

for i=1:1:row
disp(Col_Sum_alt(i,:))
end

for k=1:1:row
for i=1:1:Alt_row
   for j=1:1:Alt_row
      norm_mat_alt(j,i,k)= alt(j,i,k)/Col_Sum_alt(k,i);
   end
end
end
disp('Normalized Alternative Matrix')
for i=1:1:row
   disp(norm_mat_alt(:,:,i))
end
```

```
**************************************************
 % Take average of row to calculate priority %

for k=1:1:row
for i=1:1:Alt_row
   row_sum = 0.0;
      for j=1:1:Alt_row
            row_sum = row_sum + norm_mat_alt(i,j,k);
      end
   prio_mat_alt(k,i)= row_sum/Alt_row;
end
end

disp('Local Priority Vectors Of Alternative')
for i=1:1:row
  disp(prio_mat_alt(i,:))
end
*******************************************************
% Calculation of CR of each pair wise comparison matrix of alternatives %

for i=1:1:row
   lambda_max_alt=0.0;
   for j=1:1:Alt_row
   lambda_max_alt= lambda_max_alt + Col_Sum_alt(i,j)* prio_mat_alt(i,j);
   end
   lambda_max_new(i)= lambda_max_alt;
end
for i=1:1:row
CI_alt(i) = (lambda_max_new(i) - Alt_row) /(Alt_row -1);
end
disp('Consistency Index of Pairwise Comparion Matrix of Alternatives')
disp(CI_alt)

***************************************************
% Random Consistency Index Formula%

RI=1.98*(Alt_row-2)/Alt_row;
disp('Random Consistency Index')
disp(RI);
for i=1:1:row
CR_alt(i)=CI_alt(i)/RI;
end
disp('Consistency Ratio of Pairwise Comparison Matrix of Alternatives')
disp(CR_alt)

*****************************************************
%disp('Global Priority Vectors Of Alternative')

for i=1:1:Alt_row
   Prio_Sum =0.0;
  for j=1:1:row
   Prio_Sum = Prio_Sum + prio_mat(j)* prio_mat_alt(j,i);
  end
  global_prio(i)=Prio_Sum;
end
disp('Rank of Alternativies:')
sort_global_prio = sortrows(global_prio);
```

```
for i=1:1:Alt_row
      for j=1:1:Alt_row
if sort_global_prio(i)==global_prio(j)
   rank(j)= i;
end
   end
end
for i=1:1:Alt_row
disp(['Alternative : ', num2str(i),' Global Priority: ',num2str(global_prio(i)), ' Rank : ',num2str(rank(i))])
end
toc;
```

A.3 MATLAB Code for Extent Fuzzy AHP

```
*****************************************************************************
                    This code is developed on algorithm proposed by D.Y.Chang for Extent Fuzzy AHP

*****************************************************************************
tic;
clc;
clear % is must required to clean workspace data %
row = input('Enter no of rows or columns :');
for i = 1:1:row
  for j= 1:1:row

    if i==j
     a(i,j,1)=1;
     a(i,j,2)=1;
     a(i,j,3)=1;
        elseif (i>j)
           for k=1:1:3
              n=4-k;
                a(i,j,k)= 1/a(j,i,n) ;
                end;
    else
      a1= Extent_Fuzzy_AHP_Linguistic();
      a(i,j,1)=a1(1);
       a(i,j,2)=a1(2);
       a(i,j,3)=a1(3);
     end
  end
  end
sum_u=0.0;
sum_m=0.0;
sum_l=0.0;
for i=1:1:row
   for j=1:1:row
         sum_u=sum_u+a(i,j,3);
         sum_m=sum_m+a(i,j,2);
         sum_l=sum_l+a(i,j,1);

   end
end
disp('Sum of L M U:')
disp(sum_l)
disp(sum_m)
disp(sum_u)
```

```
**************************************************
%Creation of Cell Array %

C=cell(row,3);
for i=1:1:row
   sum_u1=0.0;
   sum_m1=0.0;
   sum_l1=0.0;
   for j=1:1:row
         sum_u1=sum_u1+ a(i,j,3);
         sum_m1=sum_m1+a(i,j,2);
         sum_l1=sum_l1+a(i,j,1);

   end
   sum_l2(i)=sum_l1;
   sum_m2(i)=sum_m1;
   sum_u2(i)=sum_u1;
   C(i,:)={sum_l2(i) sum_m2(i) sum_u2(i)};

end
disp('Extent Fuzzy Synthetic Value')
disp(C)

**************************************************
% Calculation of fuzzy synthetic extent value%

for i=1:1:row
         S(i,1)= sum_l2(i)/sum_u;
         S(i,2)=sum_m2(i)/sum_m;
         S(i,3)=sum_u2(i)/sum_l;
end
disp(S)

**************************************************
% Calculation of Degree Possibility%

for j=1:1:row
   for i=1:1:row
  if i==j
     Val(i)=196;
  else
  Val(i)= Degree_Possibility(S(j,1),S(j,2),S(j,3),S(i,1),S(i,2),S(i,3));
  end
  end
min_pos(j) = min(Val);
end
% Calculate Priority Vector%
disp('Local Priority Vector:')
disp(min_pos)

**************************************************
%Calculate Normalized Priorirty Vector%

norm_sum=0.0;
for i=1:1:row
 norm_sum=norm_sum+min_pos(i);
end
for i=1:1:row
```

```
    norm_priority(i) = min_pos(i)/norm_sum;
end
disp('Normalized Priority')
disp(norm_priority)
toc;
```

A.4 MATLAB Code for Degree of possibility

```
function Poss = Degree_Possibility(l2,m2,u2,l1,m1,u1)
%UNTITLED Summary of this function goes here
%   Detailed explanation goes here
if m2>=m1
    Poss=1;
elseif l1>=u2
    Poss=0;
else
    deno=(m2-u2)-(m1-l1);
    Poss= (l1-u2)/deno;
end
```

A.5 MATLAB Code for Extent Fuzzy AHP(linguistic variables)

```
function fuzzy_val  = Extent_Fuzzy_AHP_Linguistic()
%UNTITLED2 Summary of this function goes here
%   Detailed explanation goes here
% Declaration of TFN
Y=input('Enter Yes or No if you want to flip comparison:','s');
S = input('Enter any of the following linguistic terms : absolute,very strong,fairly strong,weak,equal:','s');
if  strcmp(Y,'No')==1
switch lower(S)
   case 'absolute'
     fuzzy_val = [3.5;4;4.5];
   case 'very strong'
     fuzzy_val = [2.5;3;3.5];
   case 'fairly strong'
     fuzzy_val = [1.5;2;2.5];
   case 'weak'
     fuzzy_val = [.67;1;1.5];
    case 'equal'
     fuzzy_val = [1;1;1];
   otherwise
     disp('Unknown Linguistic Term.')
end
else
    switch lower(S)
   case 'absolute'
      fuzzy_val = [0.2222;0.25;0.2857];
   case 'very strong'
     fuzzy_val = [0.2857;0.3333;0.4];
   case 'fairly strong'
     fuzzy_val = [0.4;0.5;0.6666];
   case 'weak'
     fuzzy_val = [0.6666;1;1.4925];

    case 'equal'
     fuzzy_val = [1;1;1];
   otherwise
```

```
    disp('Unknown Linguistic Term.')
  end
end
```

A.6 MATLAB code for Fuzzy Alpha Cut Method

```
*******************************************************************************
                    This computer code is developed for fuzzy alpha-cut method

*******************************************************************************

function a_alpha_beta = Fuzzy_Alpha_Cut()
%UNTITLED2 Summary of this function goes here
%   Detailed explanation goes here
% Declaration of TFN
S = input('Enter any of the following linguistic terms : absolute,very strong,fairly strong,weak,equal:','s');
switch lower(S)
  case 'absolute'
    fuzzy_val = [3.5;4;4.5];
  case 'very strong'
    fuzzy_val = [2.5;3;3.5];
  case 'fairly strong'
    fuzzy_val = [1.5;2;2.5];
  case 'weak'
    fuzzy_val = [.67;1;1.5];
   case 'equal'
    fuzzy_val = [1;1;1];
  otherwise
    disp('Unknown Linguistic Term.')
end
% Defuzzification of fuzzy value by fuzzy alpha cut %
%alpha = input('Enter fuzzy alpha-cut value : ');
%beta=input('Enter degree of satisfaction value : ');
alpha=0.5;
disp(fuzzy_val(2))
a_alpha_lower = (fuzzy_val(2)-fuzzy_val(1))* alpha + fuzzy_val(1);
a_alpha_upper = fuzzy_val(3)-(fuzzy_val(2) -fuzzy_val(3))* alpha;
%a_alpha_beta = (1-beta)* a_alpha_lower + beta * a_alpha_upper;

a_alpha_beta (1) =(a_alpha_lower);
a_alpha_beta (2)=(a_alpha_upper);
```

A.7 MATLAB Code for Order Allocation to selected suppliers by genetic algorithm-I

```
*******************************************************************************

Allocation of order to selected supplier/s – An example of single objective constrained optimization

*******************************************************************************

% GA FITNESS FUNCTION %
```

```
function y = simple_fitness_supplier(x)
y=-.30561*x(1)-.38463*x(2)-.30977*x(3);

****************************************************************************

% CONSTRAINT OPTIMIZATION %
function [c,ceq]=constraint_supplier(x)
c=[30*x(1)+60*x(2)+35*x(3)-65000;0.01*x(1)+0.02*x(2)+0.04*x(3)-48];
ceq=x(1)+x(2)+x(3)-1200;

****************************************************************************
clc;
objectiveFunction=@simple_fitness_supplier;
nvars=3;
LB=[0 0 0];
UB=[650 650 550];
constraintFunction=@constraint_supplier;
options=gaoptimset('PopulationSize',20,'CrossoverFraction',0.8,'MutationFcn',{@mutationadaptfeasible,0.05});
options=gaoptimset(options,'PlotFcns',{@gaplotbestf,@gaplotdistance, @gaplotrange,@gaplotbestindiv
    },'Display','iter','Generations',60);
[x,fval]=ga(objectiveFunction,nvars,[ ],[ ],[ ],[ ],LB,UB,constraintFunction,options)
```

A.8 MATLAB Code for Order Allocation to selected suppliers by genetic algorithm-II

```
****************************************************************************
Order allocation to selected supplier/s – An example of multi-objective genetic algorithm (MOGA)
****************************************************************************
function y = supplier_selection_multiobjective_fitness(x)
y(1)= 3859 * x(1)+ 3850 * x(2)+ 3851 * x(3);
y(2)= -.5654 * x(1)-.5024 * x(2)-.2033 * x(3) ;
y(3)= 0.1 * x(1) + 0.15 * x(2) + 0.2 * x(3);
y(4)= 0.2 * x(1) + 0.25 * x(3) + 0.3 * x(3);
y(5)=0.15 * x(1) + 0.2 * x(2) + 0.2 * x(3);

****************************************************************************
clc;
tic;
FitnessFunction=@supplier_selection_multiobjective_fitness;
numberOfVariables=3;
A=[2760 2750 2749]; b=[28000000];
Aeq=[1 1 1];beq=[9900];
lb=[0 0 0];
ub=[4000 3000 3000];

options=gaoptimset('PlotFcns',{@gaplotpareto});
options=gaoptimset(options,'PopulationSize',80,'HybridFcn',[],'CrossoverFraction',0.85,'CrossoverFcn',@crossoverarithmetic,'M
utationFcn',{@mutationadaptfeasible,0.5});
options = gaoptimset(options,'DistanceMeasureFcn',{@distancecrowding,'genotype'});
options = gaoptimset(options,'ParetoFraction',0.5,'Display','iter');
%options = gaoptimset(options,'PopulationSize',20);
[x,fval,exitFlag,Output,population,scores]=gamultiobj(FitnessFunction,numberOfVariables,A,b,Aeq,beq,lb,ub,options);
display(scores)
display(population)
display(fval)
fprintf('The number of points on the Pareto front was: %d\n', size(x,1));
fprintf('The average distance measure of the solutions on the Pareto front was: %g\n', Output.averagedistance);
fprintf('The spread measure of the Pareto front was: %g\n', Output.spread);
disp(['Elapsed time to solve multi-objective GA is ', num2str(toc)]);
```

A.9 MATLAB Code for Singular Value Decomposition method for fuzzy rule base reduction method

```
******************************************************************************
This computer code is developed on SVD method to reduce fuzzy rule base
******************************************************************************
function Z = Single_Value_Decomposition()
%UNTITLED Summary of this function goes here
%   Detailed explanation goes here
clc
A=[1.2857 1.2857 4 4;1.2857 4 4 6;4 4 6 6;4 6 6 8.7143] %stage-1 sustainable
%supplier selection
%A=[-10 -7.5 -5 -2.5 0;-7.5 -5 -2.5 0 2.5;-5 -2.5 0 2.5 5;-2.5 0 2.5 5 7.5;0 2.5 5 7.5 10]
%A=a;
%A=[0.8 0.8 2.5 4.5;0.8 2.5 4.5 5.5;2.5 4.5 5.5 7.5;4.5 5.5 7.5 9.2]
rank_A=rank(A);
disp(rank_A)
[U S V]=svd(A);
disp('U matrix')
disp(U)
disp('V matrix')
disp(V)
disp('Diagonal Matrix of Singular Value')
disp(S)
disp('Rank of Diagonal Matrix of Singular Value')
K=rank(S);
disp(K)
if (K>=2)
  K=2;
  disp('2 input is considered')
end
M=size(U);
N=M(2);
B=size(U);
l=B(1);
UR=U(:,1:K);
UD=U(:,K+1:N);
Cu=blkdiag(sum(UR(:,1)),sum(UR(:,2)),1);%To form the diagonal matrix
disp(Cu)
UR(1,K+1)=0;
U1= UR*Cu;
min_U1=sort(U1);
Col_U1=size(U1);

if (min_U1>= -1)
  delta=1;

else
  delta= 1/mod(min_U1);

end

for i=1:1:K
  for j=1:1:K

    if (i==j)
```

```
        stoch_matrix(i,j)=1+delta;
      else
      stoch_matrix(i,j)=1;
      end
   end

end

final_stoch_matrix =(1/(2+delta))*stoch_matrix;
U1_temp=U1(:,1:K);
U2=U1_temp*final_stoch_matrix;
disp('Prototypical Value')
disp(U2)
B=size(U2);
M=B(1);
************************************************
 %Prototypical Membership Value for FIRST Input%

EU=[U2(1,:);U2(M,:)];
%EU(K,1)=0;
EU_inv=inv(EU);
disp(EU_inv)
U3=U2*EU_inv;
disp('Membership Value for first input:')
disp(U3)


****************************************************
 %Overlapping Membership Function: Matrix V%

%V=V';
VR=V(:,1:K);
MV=size(V);
NV=MV(2);
VD=V(:,K+1:NV);
Cv=blkdiag(sum(VR(:,1)),sum(VR(:,2)),1);%To form the diagonal matrix
VR(1,3)=0;
V1= VR*Cv;
min_V1=sort(V1);
if (min_V1>= -1)
   deltaV=1;

else
   deltaV= 1/mod(min_V1);

end

for i=1:1:2
  for j=1:1:2

      if (i==j)
         stoch_matrixV(i,j)=1+deltaV;
      else
      stoch_matrixV(i,j)=1;
      end
   end

end
```

```
final_stoch_matrixV =(1/(2+delta))*stoch_matrixV;
V1_temp=V1(:,1:2);
V2=V1_temp*final_stoch_matrixV;

*************************************************************
%Prototypical Membership Value: FOR SECOND INPUT%

BV=size(V2);
V=BV(1);
EV=[V2(1,:);V2(V,:)];
EV_inv=inv(EV);
V3=V2*EV_inv;
disp('Membership Value for second input:')
disp(V3)

**********************************************************
% Reduced matrix of rule consequent values%

disp('Inverse of Eu')
inv(EU_inv)
disp('Inverse of Du')
inv(final_stoch_matrix)
disp('Inverse of Cu')
Cu1=Cu(1:2,1:2);
inv(Cu1)
S(1:2,1:2)
disp('Inverse of Cv')
Cv1=Cv(1:2,1:2);
inv(Cv1)
disp('Inverse of Dv')
inv(final_stoch_matrixV)
disp('Inverse of Ev')
inv(EV_inv')
Z=inv(EU_inv)*inv(final_stoch_matrix)*inv(Cu1)*S(1:2,1:2)*inv(Cv1)*inv(final_stoch_matrixV)*inv(EV_inv');

end
```

A.10 VB.NET Code for database connection used to prepare DSS

```
*********************************************************
A simple VB.NET code to connect SQL SERVER with front end with ADO
*********************************************************
Imports System.Data
Imports System.Data.SqlClient
Imports System.IO
Imports System.Data.Common
Partial Class _Default
   Inherits System.Web.UI.Page
   Private connect As String
   Private price(1000) As Double
   Private temp1(100) As Double
   Dim _count As Integer

   Protected Sub Page_Load(ByVal sender As Object, ByVal e As System.EventArgs) Handles Me.Load
      Dim cl As New Class_connect
      connect = cl.init_db_con
      If Page.IsPostBack = False Then
         Button1.Text = "Load Data"
      End If
      If Page.IsPostBack = True Then
         Button1.Text = "Save"
```

```
        Button1.Enabled = False
        Dim con As New SqlConnection(connect)
        con.Open()
        Dim sql As String
        sql = "Select Name,Prod_Id,Prod_Des,Price,Quantity from Supplier where Prod_Id=1002"
        'Paramteric SQL connection
        Dim myadap As New SqlDataAdapter(sql, con)
        Dim myds As New DataSet()
        myadap.Fill(myds, "Orders")
        con.Close()
        'Only the name of the field of database is required
        DDgrdVw.DataSource = myds.Tables("Orders")
        'DDgrdVw.DataValueField = "Prod_Id"
        DDgrdVw.DataBind()

        DDgrdVw.SelectedIndex = -1
        Dim i As Integer
        'Dim j As Integer
        Dim _maxValue As Double
        Dim sum As Double
        _count = myds.Tables("Orders").Rows.Count - 1
        Dim temp(_count) As Double
        'Dim temp1(j) As Double
        ReDim temp1(_count)
        ReDim price(_count)
        For i = 0 To myds.Tables("Orders").Rows.Count - 1
            price(i) = CDbl(myds.Tables("Orders").Rows(i).Item("Price").ToString)
            temp(i) = price(i)
        Next

        temp1 = temp
        Array.Sort(temp1)
        _maxValue = temp1(_count).ToString
        sum = 0.0

        For i = 0 To _count
            temp1(i) = _maxValue - price(i)
            sum = sum + temp1(i)
        Next
        For i = 0 To _count
            temp1(i) = temp1(i) / sum

        Next

    End If

  End Sub

  Protected Sub Button1_Click(ByVal sender As Object, ByVal e As System.EventArgs) Handles Button1.Click
    Dim st As Stream = File.Open("C:\price.xls", FileMode.Create, FileAccess.Write)
    Dim bw As New StreamWriter(st)
    Dim i As Integer
    For i = 0 To _count
        Dim str1 As String
        str1 = CDbl(temp1(i).ToString)
        bw.Write(str1)
        bw.WriteLine()
    Next
    bw.Close()
  End Sub
End Class
```

Kolibri 360 GmbH
Ferdinand-Jühlke-Straße 7
99095 Erfurt, Deutschland
produktsicherheit@kolibri360.de